The Ultimate Guide

to Selective & OC Maths Reasoning

Think Smarter • Build Confidence • Master Techniques

ISBN: 978-1-7645316-1-0

First published 2026
Published by iSoft, Australia

Disclaimer:
This publication is independently produced and is not endorsed by, affiliated with, or approved by any selective school, Opportunity Class (OC) authority, or official testing body.

Every effort has been made to ensure the accuracy of the content in this book. However, errors or omissions may occasionally occur. Readers are encouraged to verify solutions independently and use this material as a learning support resource.

Welcome!

I'm really glad you're here.

This book was made especially for students preparing for Selective and OC maths, students who want to think smarter, feel more confident, and maybe even start enjoying tricky problems.

Some questions will feel easy.
Some might make you pause and think.
That's completely normal, and it is actually part of learning.

You don't have to be perfect.
You don't have to get everything right straight away.
Just stay curious, keep trying, and let your thinking grow step by step.

Take your time.
Celebrate small wins.
And don't forget, **maths reasoning is a skill you build**, not something you're born with.

You're in the right place.

Let's get started.

Contents

What Selective & OC Tests Are Really About

The Selective High School Placement Test and the OC (Opportunity Class) Test are not about how many formulas or facts you can memorise.

They are about how you think, how you solve problems, and how you stay calm under time pressure.

This Test Is About Thinking, Not Memorising

You will see questions that:

→ Look unfamiliar at first
→ Combine ideas in new ways
→ Require logic before calculation

Key Insight

That's not to trick you. It's to see how well you can reason and problem-solve.

But... some memorising does help

You don't win this test by memorising everything. You win it by thinking smart. Still, there are a few facts that are worth memorising because they:

→ Save time
→ Reduce mistakes
→ Make problems easier to recognise

What This Book Includes

✓ *Only the most useful facts.*
✓ *Smart shortcuts.*
✓ *No-calculator tricks.*

Think first. Memorise a little. Solve faster.

What makes this test different?

Feature	What It Means
Fast-paced questions	About one minute per question, so efficiency matters.
Multiple choice	Use elimination, estimation, and smart shortcuts.
No calculator	Mental maths and structure are key.
Reasoning-based	You apply what you know to unfamiliar problems.

This applies to both Selective and OC tests, although OC questions are generally slightly less complex but still strongly reasoning-focused.

How This Book Helps You Succeed

By using this book, you will learn how to:

✓ Spot what a question is really asking

✓ Choose the fastest method

✓ Eliminate wrong answers

✓ Avoid common mistakes

✓ Stay calm and confident

The Bottom Line

You won't just practise questions. You'll learn ***how to think like a selective student.***

How to Play the Test Smart

Success in tests isn't just about what you know it's about how you approach each question. Master these strategies to maximise your score and minimise stress.

The 1-Minute Rule

You have about 1 minute per question.

If you:

Know what to do → solve it

Are stuck → skip it and move on

👉 **Key Insight**

Don't let one hard question steal time from five easier ones.

Do Easy First, Hard Later

Start with questions that:

✓ Look familiar.

✓ Have short working.

✓ Feel clear after one read.

Come back to the tougher ones at the end.

The Bottom Line

Smart students don't do questions in order. They do them in strategy.

Read the Question Properly

Many mistakes come from rushing.

Before solving:

→ Read the question twice.

→ Circle or underline keywords.

→ Ask: "What exactly is it asking me to find?"

Use Multiple-Choice to Your Advantage

You don't always need full working.

Try:

- ✓ **Elimination:** Cross out answers that clearly don't make sense.
- ✓ **Estimation:** Roughly work it out to see which answer is closest.
- ✓ **Substitution:** Try the answers in the question.

Stay Calm, Stay Confident

Feeling nervous is normal. What matters is how you respond.

If a question looks hard:

→ Take a breath.

→ Break it into smaller steps.

→ Give it a go.

Your Test-Day Mindset

- ✓ I don't need to know everything.
- ✓ I just need to think clearly.
- ✓ I use smart methods.
- ✓ I keep moving.
- ✓ I try every question.

Remember

- ✓ *A careful reader beats a fast guesser.*
- ✓ *Multiple choice is a tool, not a trap.*
- ✓ *Go in with confidence. Come out with confidence.*

Essential Maths Toolkit

Think first. Memorise a little. Solve faster. This chapter distills the high-value facts worth learning by heart, facts that save time, reduce mistakes, and make questions easier to recognise.

The Smart Memorisation Principle

You don't need to memorise everything. Focus on facts that:

- ✓ *Appear often in problems.*
- ✓ *Save significant working time.*
- ✓ *Reduce calculation mistakes.*

Square Numbers

Knowing square numbers helps with **area calculations**, pattern recognition, and estimation. Recognising them instantly eliminates unnecessary working.

Number	Square
1	1
2	4
3	9
4	16
5	25
6	36
7	49
8	64
9	81
10	100

Number	Square
11	121
12	144
13	169
14	196
15	225
16	256
17	289
18	324
19	361
20	400

Fraction–Decimal–Percentage Equivalents

These conversions appear constantly. Learn them once, use them forever. Being able to switch forms instantly helps you spot correct answers without lengthy calculations.

Fraction	Decimal	Percentage
1/2	0.5	50%
1/4	0.25	25%
3/4	0.75	75%
1/5	0.2	20%
2/5	0.4	40%
3/5	0.6	60%
4/5	0.8	80%
1/10	0.1	10%

Roman Numerals

Roman numerals appear in time, dates, outlines, clock faces, and reasoning questions. Fast recognition prevents errors.

Basic Roman Numerals

Roman	Value	Roman	Value
I	1	L	50
V	5	C	100
X	10	D	500
		M	1000

Common Roman Numerals

Roman	Value
II	2
III	3
IV	4
V	5
VI	6
VII	7
VIII	8
IX	9
X	10

Roman	Value
XX	20
XXX	30
XL	40
L	50
XC	90
C	100
CD	400
D	500
CM	900

Key Rules

- ✓ Symbols usually add left to right → **VIII = 8**
- ✓ Smaller before larger subtracts → **IV = 4, IX = 9**
- ✓ **I, X, C, M** may repeat up to three times
- ✓ **V, L, D** do not repeat

Angle Facts

Angle Measures

Straight line = 180°	Around a point = 360°
Right angle = 90°	Complete angle = 360°
Acute = less than 90°	Obtuse = between 90° and 180°
Reflex = between 180° and 360°	

Triangle & Circle Facts

Property	Value
Triangle interior angles total	180°
Angle in a semicircle	90°
Angle at centre	2 × angle at circumference

Equal & Supplementary Angles

Type	Property
Vertically opposite	Equal
Corresponding	Equal
Alternate	Equal

Type	Sum
Complementary	90°
Supplementary	180°
Co-interior	180°

Area & Perimeter Formulas

***Perimeter** = distance around a shape (units: cm, m, km)*

***Area** = space inside a shape (units: cm^2, m^2, km^2)*

Shape	Perimeter	Area
Rectangle	2 × (l + w)	l × w
Square	4 × side	$side^2$
Triangle	a + b + c	(base × height) ÷ 2
Parallelogram	2 × (a + b)	base × height
Trapezium	a + b + c + d	(Temperature × height) ÷ 2
Circle	$2\pi r$ (circumference)	πr^2

Area Conversions

Conversion	Value
1 cm^2	100 mm^2
1 m^2	10,000 cm^2
1 hectare	10,000 m^2

Units & Conversions

Length

10 mm = 1 cm

100 cm = 1 m

1000 m = 1 km

Mass

1000 mg = 1 g

1000 g = 1 kg

1000 kg = 1 tonne

Capacity & Volume

1000 mL = 1 L

1000 L = 1 kL

1000 L = 1 m^3

Time

Conversion	Value
1 minute	60 seconds
1 hour	60 minutes
1 day	24 hours
1 week	7 days
1 year	365 days (366 in leap year)
1 year	52 weeks
1 decade	10 years
1 century	100 years

Useful Time Facts

→ *Minutes per day = 1440*

→ *Seconds per hour = 3600*

→ *Seconds per day = 86,400*

→ *½ hour = 30 min*

→ *¼ hour = 15 min*

→ *¾ hour = 45 min*

Speed, Distance & Time

Speed *= distance ÷ time*

Distance *= speed × time*

Temperature

***°F** = (°C × 9/5) + 32*

***°C** = (°F − 32) × 5/9*

Money

***100 cents** = $1*

***Total cost** = price × quantity*

***Change** = amount given − cost*

Divisibility Rules

These rules allow you to quickly determine if a number divides evenly without performing full division.

Divisible by	Quick Check
2	Last digit is even (0, 2, 4, 6, 8)
3	Sum of digits is divisible by 3
5	Ends in 0 or 5
9	Sum of digits is divisible by 9
10	Ends in 0

Example:

Is 57 divisible by 3?

Sum of digits: $5 + 7 = 12$. Since $12 \div 3 = 4$ (whole number), yes, 57 is divisible by 3.

Spatial & Shapes

2D Shape Names

Shape	Edges (Sides)	Vertices (Corners)
Triangle	3	3
Quadrilateral (Square, Rectangle, Parallelogram, Rhombus, Trapezoid, Kite)	4	4
Pentagon	5	5
Hexagon	6	6
Octagon	8	8
Nonagon	9	9
Decagon	10	10

3D Shapes

Shape	Faces	Edges	Vertices
Cube	6	12	8
Rectangular prism	6	12	8
Triangular prism	5	9	6
Square pyramid	5	8	5
Cylinder	3	2	0
Sphere	1 (curved)	0	0

Common Nets

Cube	6 squares
Triangular prism	2 triangles + 3 rectangles
Cylinder	2 circles + 1 rectangle
Sphere	No net (cannot be unfolded flat)

Common Creatures & Leg Counts

Creature	Legs
Ant	6
Bee	6
Fly / Mosquito	6
Butterfly / Moth	6
Grasshopper / Cricket	6
Cockroach	6
Dragonfly	6
Spider	8
Scorpion	8
Crab	10
Lobster	10

Why is this table used in maths?

*In school problem-solving (especially selective-style questions), students often use the **total number of legs and creatures** to work out how many of each type there are. Animals with fixed leg counts (like insects = 6, spiders = 8) make the maths easier to solve using equations or logical reasoning. Remembering these common leg counts can also **save time and reduce guessing** during tests.*

Quick Self-Check

Can you answer these instantly? If yes, you're using memory the smart way.

Test Yourself

1. What is 12^2?
2. What percentage is 3/4?
3. How many degrees in a triangle?
4. How many mL in 2 L?
5. What is IX in Arabic numerals?
6. Is 57 divisible by 3? Why?

👉 The Golden Rule of Memorising

Don't memorise more. Memorise smarter.

If a fact appears often, saves time, and reduces mistakes, it belongs in your memory. Everything else? Use reasoning instead.

Prime, Composite & Number Properties

Numbers have personalities. Some like to stand alone, while others love to share. If you can recognise a number's "personality," you will solve faster, eliminate wrong answers, and avoid unnecessary working. This chapter gives you high-value shortcuts for selective school questions.

Your Memory Anchor

Divide by one and itself? Prime.
Divide by one, itself, and more? Composite every time.
1 is neither! 2 is prime!

The Three Number Types You Must Know

Prime Numbers

A prime number can be divided by exactly two numbers: 1 and itself.

Examples: 2, 3, 5, 7, 11, 13, 17...

From the rhyme: "Divide by one and itself? Prime."

Composite Numbers

A composite number can be divided by more than two numbers.

Examples

4 → 1, 2, 4
6 → 1, 2, 3, 6
8, 9, 10, 12...

From the rhyme: "Divide by one, itself, and more? Composite every time."

The Special Numbers: 1 and 2

- *1 is neither prime nor composite (it has only one factor).*
- *2 is prime (it has exactly two factors: 1 and 2).*

From the rhyme: "1 is neither! 2 is prime!"

Selective tests love these traps.

What These Questions Look Like

You might be asked:

- *Which of the following is prime?*
- *How many factors does 24 have?*
- *Which is the largest composite number less than 40?*
- *Which number is divisible by both 3 and 5?*

If you know number types, these become quick wins.

FAST CHECK (5 seconds)

Before doing any long working, eliminate quickly:

Not prime if:

- *It is even (except 2)*
- *It ends in 0 or 5 (except 5)*
- *The sum of digits is divisible by 3 (except 3)*

If any of these are true → **composite**.

This instantly removes most wrong options.

Multiple-Choice Shortcut: "Prime Hunt"

When asked: Which number is prime?

- *Cross out even numbers (except 2)*
- *Cross out numbers ending in 0 or 5 (except 5)*
- *Cross out numbers divisible by 3*

This can turn a 30-second question into 5 seconds.

Worked Example

Question: Which of the following numbers is prime?
A) 21 B) 27 C) 29 D) 33 E) 35

Think smart:

21 → divisible by 3 → composite

27 → divisible by 3 → composite

29 → not even, not divisible by 3 or 5 → ✓ prime

33 → divisible by 3 → composite

35 → ends in 5 → composite

✓ Answer: C) 29

Factors: The Building Blocks

What is a factor?

A factor is a number that divides another number exactly, meaning there is no remainder.

Another easy way to think:

If two numbers multiply to make a number, they are its factors.

Example:

12 = 3 × 4
So **3** and **4** are factors of **12**.

Worked Example

Question: Let's Work Out the Factors of 12.

Steps:

12 ÷ 1 = 12 ✓ → factors: 1 and 12
12 ÷ 2 = 6 ✓ → factors: 2 and 6
12 ÷ 3 = 4 ✓ → factors: 3 and 4

After this, the pairs repeat, so we **stop**.

✓ **Answer:** Final factors of 12: 1, 2, 3, 4, 6, 12

What Are Common Factors?

A common factor is a factor that two or more numbers share.

Factors of 12 → 1, 2, 3, 4, 6, 12

Factors of 18 → 1, 2, 3, 6, 9, 18

✓ 1, 2, 3, and 6 appear in both lists.

These are the common factors.

Highest Common Factor (HCF)

The Highest Common Factor (HCF) is the largest number that divides two or more numbers exactly.

To find it:

1. List factors of each number.
2. Identify numbers appearing in both lists.
3. Choose the largest one.

Factors of 12 → 1, 2, 3, 4, **6**, 12
Factors of 18 → 1, 2, 3, **6**, 9, 18

✓ We already know 1, 2, 3, and 6 are common factors. The largest of these is **6**.

So, **HCF** = **6**

Why Factors Matter

Simplifying Fractions: $\frac{18}{24}$

Common factor = 6.

Divide both:

$18 \div 6 = 3$
$24 \div 6 = 4$

Result: $\frac{3}{4}$

Factors make fractions simpler and cleaner.

Factor Patterns Worth Knowing

Even Numbers

Even numbers always have at least:

- ✓ 1
- ✓ 2
- ✓ Half the number
- ✓ Itself

Example: 10 → 1, 2, 5, 10

Because they have multiple factors, even numbers (except 2) are usually composite.

Divisibility Rules = Speed

Use these to spot composites instantly:

Divisible by	Rule	Example
2	Ends in 0, 2, 4, 6, 8	48 → ends in 8 ✓
3	Digit sum divisible by 3	57 → 5+7=12 ✓
5	Ends in 0 or 5	85 → ends in 5 ✓
9	Digit sum divisible by 9	81 → 8+1=9 ✓
10	Ends in 0	120 → ends in 0 ✓

Worked Example

Question: Is 57 prime?

5 + 7 = 12 → divisible by 3 → **composite**

No long division needed.

Why This Chapter Matters

Knowing primes, composites, and factors helps with:

- ✓ Pattern questions.
- ✓ Divisibility problems.
- ✓ Largest/smallest number questions.
- ✓ Multiple-choice elimination.
- ✓ Fraction simplification.
- ✓ Probability.

This is one of the highest-value skills in the selective test.

Quick Self-Check

1. Which is prime?
   A) 15 B) 17 C) 21 D) 27
2. How many factors does 20 have?
3. What is the smallest composite number?
4. Is 49 prime or composite? Why?
5. Which number is divisible by both 3 and 5?
   A) 14 B) 20 C) 30 D) 41

Key Takeaway

Divide by one and itself? Prime.

Divide by one, itself, and more? Composite every time.

1 is neither! 2 is prime!

You don't need to factor everything.

Think smart. Eliminate fast. Solve with confidence.

PRIME & COMPOSITE NUMBERS: PROPERTIES & FAST CHECKS

PRIME NUMBERS

Exactly <u>TWO</u> factors: 1 & itself.

Examples:
2, 3, 5, 7, 11, 13, 17...

MEMORY ANCHOR

Divide by one & itself? Prime.
Divide by one, itself, & more? Composite every time.
1 is neither! 2 is prime!

COMPOSITE NUMBERS

<u>MORE</u> than two factors.

Examples:
4 (1,2,4), 6 (1,2,3,6), 8, 9, 10, 12...

SPECIAL NUMBERS (TRAPS!)

(1) 1: NEITHER (only 1 factor)

(2) 2: PRIME (exactly 2 factors: 1 & 2)

FACTORS & DIVISIBILITY (SPEED!)

FACTORS:
Building blocks, divide exactly (no remainder).
E.g., Factors of 12:
1, 2, 3, 4, 6, 12.

→

DIVISIBILITY RULES (SPOT COMPOSITES)
By 2 → ends in 0, 2, 4, 6, 8.
By 3 → digit sum divisible by 3.
By 5 → ends in 0 or 5.
By 9 → digit sum divisible by 9.

FAST CHECK (5 SECONDS)

Before long working, ELIMINATE QUICKLY! → NOT PRIME IF:
- Even (except 2)?
- Ends in 0 or 5 (except 5)?
- Digit sum divisible by 3 (except 3)?

IF <u>ANY</u> TRUE → COMPOSITE

WHY IT MATTERS

Patterns, Divisibility, Largest/Smallest, MC Elimination, Fractions, Probability.

Number & Place Value

Don't Skip This Chapter (Seriously)

This chapter can look easy. Many students skim it, and many students lose marks because of that. Selective-style maths questions often do not test difficult calculations, they test whether you truly understand how numbers are built.

Small place value mistakes cause:

- ☒ Wrong comparisons
- ☒ Incorrect rounding
- ☒ Misread decimals
- ☒ Avoidable exam errors

So slow down here. Master this once, and it will quietly improve almost every maths topic that follows.

Your Memory Anchor

Biggest place first.
Compare left to right.
Round by the next digit.
See it, don't calculate.

Why This Topic Matters

When you clearly understand place value, you can:

- ✓ Compare numbers instantly
- ✓ Spot largest or smallest values quickly
- ✓ Estimate confidently
- ✓ Eliminate wrong answers without heavy working

This chapter is about learning to **see numbers clearly**, not just calculate with them. And maths becomes much more enjoyable when things feel obvious.

What These Questions Usually Look Like

You'll often see questions like:

- ✓ Which number is largest or smallest?
- ✓ Which number is closest to a given value?
- ✓ Which has the greatest digit in a certain place?
- ✓ Which number rounds to a particular value?

These are thinking questions, not calculating questions. Strong students do not rush to calculate, they analyse the structure first.

Place Value Comes First

Every number is built from places.

Place	**Example Value**
Ten-thousands	**50,000**
Thousands	**5,000**
Hundreds	**500**
Tens	**50**
Ones	**5**

Comparing Numbers Quickly

What are we solving?

Which number is larger or smaller?

Formula (Place Comparison)

1. Start with the biggest place value
2. Compare digits from left to right
3. Stop as soon as a difference appears

Worked Example

Question: Which is larger: **48,736 or 48,691?**

Think smart:

4 8 7 3 6

4 8 6 9 1

- Ten-thousands: both 4
- Thousands: both 8
- Hundreds: **7 vs 6**
- So, 7 > 6

✓ **Answer:** 48,736 is larger

⚠ **Visual Tip**

Do not compare the whole number, compare where the difference begins.

First Digit Wins

What are we solving?

When comparing large numbers in multiple-choice maths questions, carefully checking every digit can take unnecessary time. This method helps you quickly spot the largest or smallest number by focusing on the first digit first.

It helps you:

✓ Eliminate wrong options quickly

✓ Save time for harder questions

✓ Avoid unnecessary calculation

✓ Work faster and more confidently

Because place value matters most from left to right, the **first digit usually determines which number is larger or smaller**. Only check later digits if the first digits match.

Formula (Cross Method)

- ✓ Largest number → eliminate smaller leading digits
- ✓ Smallest number → eliminate larger leading digits

Worked Example

Question: Which number is greatest?

Think smart:

A) **4**,938
B) **5**,102
C) **4**,999
D) **4**,850
E) **4**,976

Only one number starts with **5**.

✓ **Answer: 5,102**

Closest-To Questions

What are we solving?

Which number is nearest a target value?

Formula (Estimation Method)

- ✓ Lock onto the target number
- ✓ Estimate each distance
- ✓ Choose the smallest gap

Worked Example

Target: **10,000**

Options:

9,800 → 200 away

10,300 → 300 away

10,050 → **50** away

✓ Closest = **10,050**

Comparing Decimals

Decimals often trick students because digits look longer or shorter.

Quick Rule

- ✓ Line numbers up at the decimal point
- ✓ Add zeros if needed
- ✓ Compare place by place

Worked Example

Compare: 0.5 vs 0.47

Write as:

0.50
0.47

5 tenths > 4 tenths.

✓ Answer: 0.5 is larger

Quick Checklist

Skill	Ask Yourself
Comparing numbers	Did I start from the biggest place?
Rounding	Did I check the next digit only?
Decimals	Did I line up decimal points?
Closest values	Did I estimate distance?
Exam shortcuts	Did I eliminate obvious wrong answers?

Try It Yourself

1. Which is largest?
   A) 3,492 B) 3,509 C) 3,487 D) 3,499
2. Which is smallest?
   A) 0.605 B) 0.65 C) 0.56 D) 0.602
3. Which rounds to 600 (nearest hundred)?
   A) 549 B) 551 C) 650 D) 601
4. Which is closest to 2,000?
   A) 1,920 B) 2,080 C) 2,150 D) 1,700
5. Order from smallest to largest:
   0.4, 0.35, 0.375, 0.5

Key Takeaway

Biggest place first. Compare smart. Don't overthink.

You're not racing to calculate. You're racing to see.

Place value may look simple, but mastering it gives you a real advantage.

NUMBER & PLACE VALUE: DON'T SKIP THIS CHAPTER (SERIOUSLY)

Master **structure** to improve almost every maths topic. See it, don't just calculate.

WHY IT MATTERS (The Hidden Trap)

Skimming leads to avoidable errors:

- ✗ Wrong comparisons
- ✗ Incorrect rounding
- ✗ Misread decimals.

THE BENEFITS (Thinking, Not Calculating)

- ✓ Compare instantly
- ✓ Spot largest/smallest quickly
- ✓ Estimate confidently
- ✓ Eliminate wrong answers fast.

YOUR MEMORY ANCHOR

- **Biggest place first.**
- **Compare left to right.**
- **Round by the next digit.**

UNDERSTANDING STRUCTURE

Place Value Table				
Ten-thousands	Thousands	Hundreds	Tens	Ones
5	0	5	5	5

← Value increases

TECHNIQUE 1: COMPARING QUICKLY

Formula (Place Comparison):

1. Start at biggest place value.
2. Compare digits left to right.
3. STOP as soon as difference appears.

X X [Y] Z
X X [W] Z ← Focus where difference begins.

SMART TACTICS & EXAM SHORTCUTS

TECHNIQUE 2: FIRST DIGIT WINS (Cross Method)

A) ~~X~~...
B) ~~X~~...
C) ~~X~~...

← Largest number? Eliminate smaller leading digits fast.

TECHNIQUE 3: CLOSEST-TO QUESTIONS (Estimation)

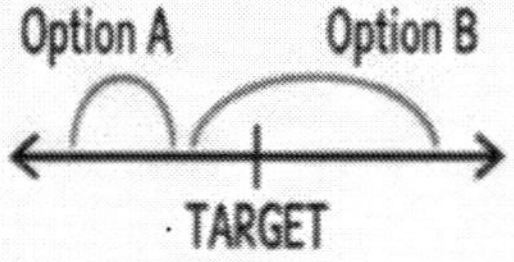

Lock onto target. Choose smallest gap.

TECHNIQUE 4: COMPARING DECIMALS (Line 'Em Up)

0.X[0]0
0.X Y Z

Line up decimal points. Add zeros to compare.

QUICK EXAM CHECKLIST

☐ Started from biggest place? ☐ Checked next digit only for rounding? ☐ Lined up decimal points? ☐ Estimated distance for 'closest'? ☐ Eliminated obvious wrong answers?

KEY TAKEAWAY: You're not racing to calculate. You're racing to SEE structure. Don't overthink it.

Fractions, Decimals & Percentages

Different Forms, Same Value

Fractions, decimals, and percentages all show **parts of a whole**. They may look different, but they express the same idea. Selective and OC tests often switch between them quickly. If you can move easily between the three, you'll work faster, make fewer mistakes, and choose answers confidently.

This chapter is about **thinking smart instead of working long**.

Your Memory Anchor

Fraction → Divide → Decimal → ×100 → Percent

This simply means:

- A fraction becomes a decimal by dividing.
- A decimal becomes a percentage by multiplying by 100.
- All three forms represent the same value.

$$\mathbf{\frac{3}{4} \rightarrow 3 \div 4 = 0.75 \rightarrow 0.75 \times 100 = 75\%}$$

What These Questions Look Like

You might see:

- ✓ What is ¾ of 40?
- ✓ Which is bigger: 0.6 or ⅝?
- ✓ What is 25% of 80?
- ✓ Which picture shows ⅔ shaded?
- ✓ If 40% of students are girls, how many girls out of 50?

They can look tricky, but the right method makes them straightforward.

The Three Simple Rules

Rule 1: "Of" Means Multiply

$$\frac{3}{4} \text{ of } 40 = \frac{3}{4} \times 40$$

Whenever you see **"of"**, think multiplication.

Rule 2: Change to What's Easiest

Don't stay stuck in one form.

- ✓ Use a **fraction** if it simplifies neatly
- ✓ Use a **decimal** if comparing or calculating quickly
- ✓ Use a **percentage** if it feels intuitive

Choose the form that makes the maths easier.

Rule 3: Cancel Before You Multiply

Make numbers smaller first before you multiply.

Smaller numbers = easier maths = fewer mistakes.

Example: $\frac{3}{4} \times 20$

Do this first: $20 \div 4 = 5$, **then** $5 \times 3 = 15$

Finding a Fraction of a Number

This appears often in discounts, recipes, word problems, and test questions.

What are we solving?

Finding a part of a number.

Formula (Divide Then Multiply)

1. Divide by the denominator.
2. Multiply by the numerator.

Info

- ***Numerator:*** *Top number*
- ***Denominator:*** *Bottom number*

Worked Example

Question: What is $\frac{5}{6}$ **of 42**?

Steps:

$42 \div 6 = 7$
$7 \times 5 = 35$

✓ **Answer:** 35

Converting Between Forms

Fraction → Decimal

Decimals are often easier for calculations, measurements, and comparisons.

What are we solving?

Writing a fraction in decimal form.

Formula (Division Method)

$$\text{Decimal} = \text{Numerator} \div \text{Denominator}$$

Example: $\frac{3}{4} \rightarrow 3 \div 4 = 0.75$

Decimal → Percentage

Percentages are widely used for scores, discounts, interest rates, and data comparisons.

What are we solving?

Expressing a decimal as a percentage.

Formula (Scaling Method)

$$\text{Percent} = \text{Decimal} \times 100$$

Example: $0.6 \times 100 = 60\%$

Move the decimal 2 places right.

Percentage → Decimal

Sometimes percentages are given, but calculations require decimals.

What are we solving?

Writing a percentage as a decimal.

Formula

$$\text{Decimal} = \text{Percent} \div 100$$

Example: $25\% \div 100 = 0.25$

Move the decimal 2 places left

Percentage → Fraction (easy ones)

Some percentages convert easily into simple fractions:

$$50\% = \frac{1}{2}$$

$$25\% = \frac{1}{4}$$

$$20\% = \frac{1}{5}$$

$$75\% = \frac{3}{4}$$

Using these makes mental maths faster.

Helpful Values to Remember

Fraction	Decimal	Percentage
$\frac{1}{2}$	0.5	50%
$\frac{1}{4}$	0.25	25%
$\frac{3}{4}$	0.75	75%
$\frac{1}{5}$	0.2	20%
$\frac{2}{5}$	0.4	40%
$\frac{3}{5}$	0.6	60%
$\frac{4}{5}$	0.8	80%
$\frac{1}{10}$	0.1	10%

These common values save time in exams.

Worked Example

Question: Which is bigger: $\frac{3}{4}$ **or 0.7?**

Think smart:

Convert the fraction:

$$\frac{3}{4} = 0.75$$

Compare:

0.75 vs **0.7**

✓ **Answer:** $\frac{3}{4}$ is bigger

Comparing Fractions: Cross-Multiply

When **denominators differ**, **cross-multiplying** allows quick comparison without converting.

What are we solving?

Comparing two fractions efficiently.

Formula (Cross Method)

For $\frac{a}{b}$ and $\frac{c}{d}$

Multiply a × d and c × b.

💡 The **larger** result indicates the **larger fraction.**

Worked Example

Question: Compare $\frac{2}{3}$ and $\frac{3}{5}$:

Think smart:

a × d and c × b.

2 × 5 = **10**
3 × 3 = 9

✓ **Answer:** $\frac{2}{3}$ is bigger.

Question: Compare $\frac{3}{4}$ and $\frac{2}{5}$:

Think smart:

3 × 5 = 15
2 × 4 = 8

✓ **Answer:** $\frac{3}{4}$ is bigger.

Percentages Made Easy

Certain percentages appear frequently and convert neatly into fractions.

Quick Facts

10% → divide by 10
20% → double 10%
25% → divide by 4
50% → divide by 2
75% → 50% + 25%

Patterns reduce working.

Worked Example

Question: What is 25% of 80?

Step 1: Remember — "of" means multiply

25% of 80 = 25% × 80

Step 2: Change 25% to an easy fraction

$25\% = \frac{25}{100} = \frac{1}{4}$

Step 3:

$\frac{1}{4} \times 80$

Divide first to keep numbers small

80 ÷ 4 = 20

✓ **Answer:** 20

Worked Example

Question: 40% of 50 students are girls. How many girls?

Step 1: Remember — "of" means multiply

$40\% \text{ of } 50 = 40\% \times 50$

Step 2: Change 40% to an easy fraction

$40\% = \frac{40}{100} = \frac{2}{5}$

Step 3:

$\frac{2}{5} \times 50$

Divide first to keep numbers small

$50 \div 5 = 10$

Now multiply

$2 \times 10 =$ **20**

✓ **Answer:** 20

Quick Checklist Summary

Situation	**What To Do**
Fraction of a number	Divide, then multiply
Fraction → Decimal	Divide numerator by denominator
Decimal → Percent	Multiply by 100
Percent → Decimal	Divide by 100
Compare fractions	Cross-multiply
Percentage of number	Use friendly fraction or 10% rule

Why This Chapter Matters

You'll see these ideas in:

- ✓ Word problems
- ✓ Graphs and diagrams
- ✓ Probability
- ✓ Ratios
- ✓ Data interpretation

Understanding the connection between fractions, decimals, and percentages makes maths clearer and builds confidence.

Quick Self-Check

1. What is 3/5 of 45?
2. Which is bigger: 0.6 or 5/8?
3. What is 20% of 90?
4. A pizza has 12 slices. If 9 are eaten, what fraction is left?
5. Which is closest to 1/2?
   A) 0.45 B) 0.6 C) 0.52 D) 0.4

Key Takeaway

Fraction → Divide → Decimal → ×100 → Percent

Fractions, decimals, and percentages are not different ideas.

They are just different ways to show the same part of a whole.

FRACTIONS, DECIMALS & PERCENTAGES: DIFFERENT FORMS, SAME VALUE

Parts of a Whole – Expressing the Same Idea

THE THREE SIMPLE RULES

1. "Of" Means Multiply
   (e.g., ¼ of 40 = ¼ × 40)
2. Change to What's Easiest
   (Fraction, Decimal, or Percent)
3. Cancel Before You Multiply
   (Smaller numbers = easier maths)

FINDING A FRACTION OF A NUMBER

Formula: (Divide Then Multiply)

1. Divide by Denominator (bottom)
2. Multiply by Numerator (top)

MEMORY ANCHOR

Fraction —(Divide)→ Decimal —(×100)→ Percent

$\frac{3}{4} \rightarrow 3 \div 4 = 0.75 \rightarrow 0.75 \times 100 = 75\%$

CONVERTING BETWEEN FORMS

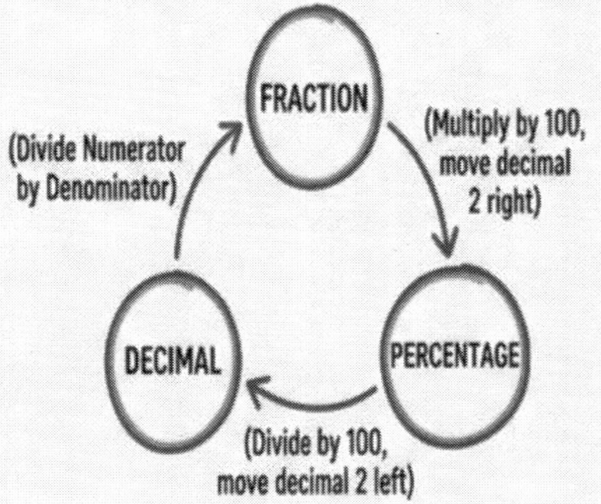

HELPFUL VALUES		
$\frac{1}{2}$	= 0.5	= 50%
$\frac{1}{4}$	= 0.25	= 25%
$\frac{3}{4}$	= 0.75	= 75%
$\frac{1}{5}$	= 0.2	= 20%
$\frac{2}{5}$	= 0.4	= 40%
$\frac{3}{5}$	= 0.6	= 60%
$\frac{5}{5}$	= 0.8	= 80%
$\frac{11}{0}$	= 0.1	= 10%

KEY TAKEAWAY

Fractions, decimals, and percentages are just different ways to show the same part of a whole. Master conversions for speed and confidence.

COMPARING FRACTIONS: CROSS-MULTIPLY

$\frac{a}{b}$ ↗ $a \times d$

$\frac{c}{d}$ ↘ $c \times b$

Compare results:
Larger result = Larger fraction

PERCENTAGES MADE EASY (QUICK FACTS)

10% → divide by 10
20% → double 10%
25% → divide by 4
50% → divide by 2
75% → 50% + 25%

Ratio & Proportion

Seeing Parts, Sharing Smartly

Sharing made simple when you focus on parts, everything becomes clear. In this chapter you'll learn a repeatable method to share, compare, and solve ratio word problems with confidence.

Your Memory Anchor

BALANCE *keeps it fair.*

If one side grows, match it there.

CROSS *goes diagonal.*

Multiply across, quick and logical.

PARTS *means ALL, ONE, MULTIPLY.*

Add the parts, find one, then apply.

A ratio compares amounts. It shows how much of one thing there is compared to another. For example, a ratio of **2 : 3** means for every 2 of one thing, there are 3 of another.

In ratio problems, it helps to stop thinking about the actual numbers first and instead think about **parts**. Once you know what **one part** is worth, everything becomes easy.

Ratio questions often look different, but most of them are the same problem wearing a different costume. The secret is to stop chasing the numbers and start seeing parts.

Where We Use Ratios

Ratios appear everywhere:

- ✓ Recipes (2 cups flour : 1 cup sugar)
- ✓ Sharing food or money
- ✓ Map scales
- ✓ Mixing colours
- ✓ Sports statistics

Ratios help keep things balanced.

How Ratios Are Written

All of these mean the same:

- ✓ 2 to 3
- ✓ 2 : 3 ← most common
- ✓ 2/3

Sharing when the total is given

Nai and Mika found a bag with 40 stickers. They decide to share them fairly based on how much they helped clean the classroom. They agree on the ratio Nai : Mika = 3 : 5.

What are we solving?

We need to split 40 into two amounts that keep the relationship 3 parts to 5 parts. This is a PARTS method problem (sharing a total).

Formula

PARTS → ALL → ONE → MULTIPLY

Worked Example

Step 1 (ALL): Add the parts

$3 + 5 = 8$
This tells us how many equal parts the total is split into.

Step 2 (ONE): Find one part

$40 \div 8 = 5$
Each part is worth 5 stickers.

Step 3 (MULTIPLY): Find each share

Nai = $3 \times 5 = 15$
Mika = $5 \times 5 = 25$
Multiply each person's parts by the value of one part.

✓ **Answer:** Nai = 15, Mika = 25

Finding one group when the total is given

A fruit box has apples and oranges in the ratio 2 : 3. The box has 30 fruits altogether. You want to know how many are apples.

What are we solving?

We're sharing a total into ratio parts again. Still the **PARTS method**.

Formula

PARTS → ALL → ONE → MULTIPLY

Worked example

Step 1 (ALL): Add the parts

2 + 3 = 5
Total parts in the box.

Step 2 (ONE): Find one part

30 ÷ 5 = 6
One part equals 6 fruits.

Step 3 (MULTIPLY): Find apples

Apples = 2 × 6 = **12**
Apples are 2 parts.

✓ **Answer:** 12 apples

When the total is NOT given

In a school photo, the ratio of boys to girls is 5 : 7. The teacher counts 35 boys. How many girls are there?

What are we solving?

We know one group but not the total. We scale from known parts. This is still the **PARTS scaling method**.

Formula

PARTS thinking → Find ONE part first, then multiply.

Worked example

Step 1: Match the given number to its parts

Boys = 5 parts = 35
We connect the real number to the ratio.

Step 2 (ONE): Find one part

1 part = 35 ÷ 5 = 7
Divide by the number of parts you were given.

Step 3 (MULTIPLY): Find girls

Girls = 7 parts = 7 × 7 = 49
Girls are 7 parts.

✓ **Answer:** 49 girls

Keeping a mixture balanced

A drink is mixed in the ratio Water : Syrup = 4 : 1. You already poured 200 mL of water. How much syrup should you add?

What are we solving?

We must keep the taste consistent by keeping the ratio balanced. This uses the **PARTS method (mixture scaling)**.

Formula

PARTS → ONE part → Multiply remaining parts

Worked example

Step 1: Water = 4 parts = 200 mL

Step 2 (ONE): 200 ÷ 4 = 50 mL

Step 3: Syrup = 1 part = 50 mL

✓ **Answer:** 50 mL syrup

Simplifying ratios

A class has 12 boys and 18 girls. The teacher wants the ratio in simplest form.

What are we solving?

We want an equivalent ratio that keeps balance but uses smaller numbers. This uses the **BALANCE method** (divide both equally).

Formula

BALANCE → Divide both numbers by the biggest common divider.

Worked example

Step 1: Start with the ratio

12 : 18

Step 2: Find the biggest number that divides both

Biggest common divider = 6.
This is sometimes called the HCF (Highest Common Factor) but think "biggest divider."

Step 3: Divide both numbers

12 ÷ 6 = 2
18 ÷ 6 = 3
Dividing both keeps the ratio balanced.

✓ **Answer:** 2 : 3

Ratio to fraction

A jar has red and blue marbles in the ratio 3 : 1. Find the fraction that are red.

What are we solving?

Turning ratio parts into a fraction of the whole. This is mainly **PARTS thinking**.

Formula

Total parts = a + b

Fraction = part ÷ total parts

Worked example

Step 1: Add the parts to find the whole

3 + 1 = 4
There are 4 equal parts total.

Step 2: Put the red parts over the total

Red fraction = 3/4
Red is 3 out of 4 parts.

✓ **Answer:** 3/4

Quick Checklist

If Question Says...	Use This Method
Share / Divide / Total given	ALL → ONE → MULTIPLY
One value given only	Find ONE part first
Simplest form	Divide both numbers
Missing number ratio	Cross multiply or scale
Recipe / mixture / scaling	Keep ratio balanced
Fraction asked	Part ÷ total parts

Try It Yourself

1. Divide 72 in the ratio 5 : 3.
2. Share 56 marbles in the ratio 2 : 6.
3. Boys : Girls = 4 : 9. There are 36 girls. How many boys?
4. Simplify 20 : 30.
5. Simplify 45 : 60.
6. 3 : 7 = 12 : x
7. 5 : 2 = x : 8
8. A recipe uses milk : water = 3 : 2. If 600 mL milk is used, how much water?

9. Ratio 4 : 1 — fraction of the smaller part?

10. Ratio 5 : 3 — fraction of the larger part?

Key Takeaway

ALL in ONE, MULTIPLY.

Ratios aren't about big numbers. They're about seeing the parts and scaling smartly.

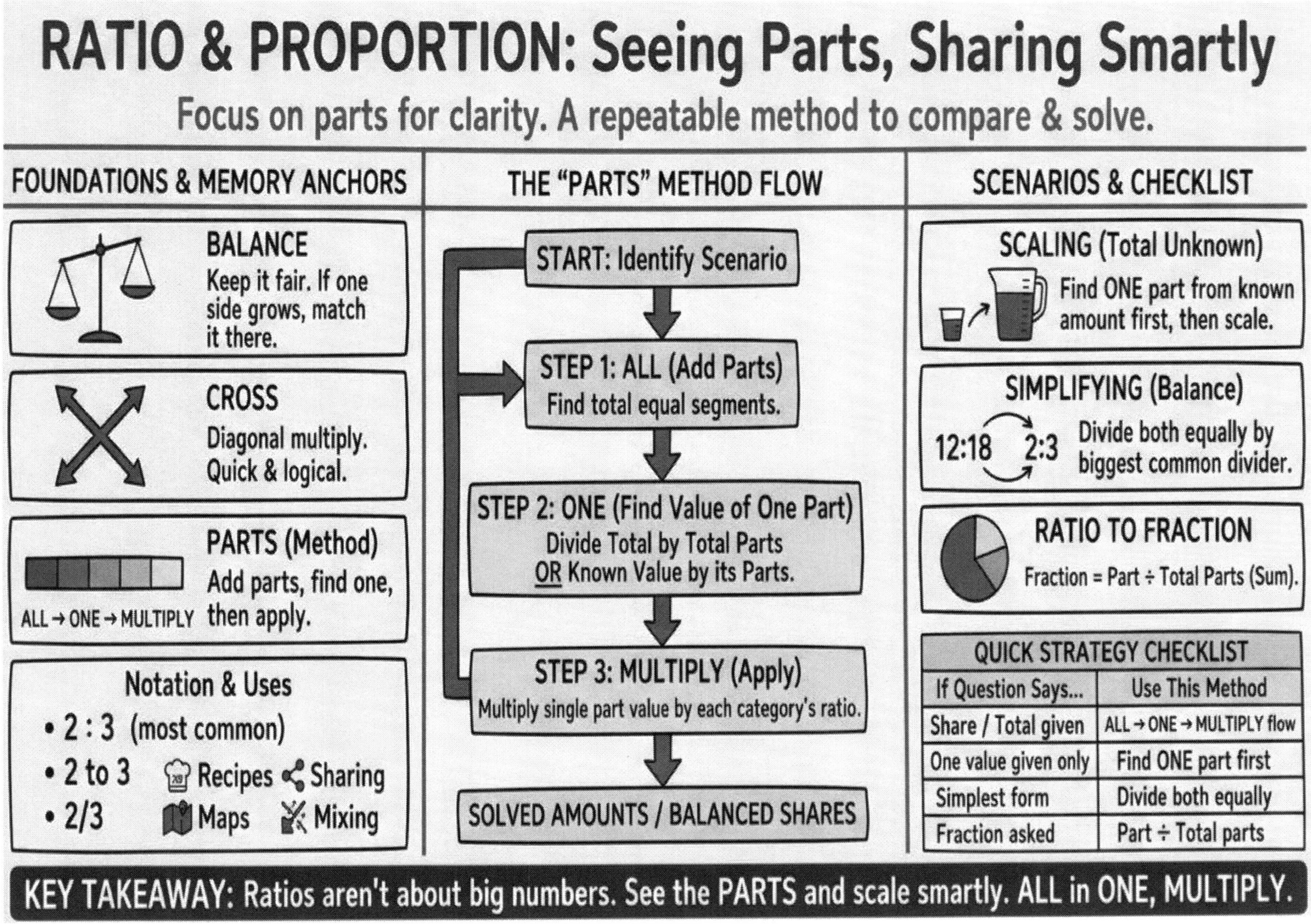
RATIO & PROPORTION: Seeing Parts, Sharing Smartly
Focus on parts for clarity. A repeatable method to compare & solve.
FOUNDATIONS & MEMORY ANCHORS
BALANCE
Keep it fair. If one side grows, match it there.
CROSS
Diagonal multiply. Quick & logical.
PARTS (Method)
Add parts, find one, then apply.
ALL → ONE → MULTIPLY
Notation & Uses
• 2 : 3 (most common)
• 2 to 3
• 2/3
Recipes
Sharing
Maps
Mixing
THE "PARTS" METHOD FLOW
START: Identify Scenario
STEP 1: ALL (Add Parts)
Find total equal segments.
STEP 2: ONE (Find Value of One Part)
Divide Total by Total Parts
OR Known Value by its Parts.
STEP 3: MULTIPLY (Apply)
Multiply single part value by each category's ratio.
SOLVED AMOUNTS / BALANCED SHARES
SCENARIOS & CHECKLIST
SCALING (Total Unknown)
Find ONE part from known amount first, then scale.
SIMPLIFYING (Balance)
12:18
2:3
Divide both equally by biggest common divider.
RATIO TO FRACTION
Fraction = Part ÷ Total Parts (Sum).
QUICK STRATEGY CHECKLIST
If Question Says... | Use This Method
Share / Total given | ALL → ONE → MULTIPLY flow
One value given only | Find ONE part first
Simplest form | Divide both equally
Fraction asked | Part ÷ Total parts
KEY TAKEAWAY: Ratios aren't about big numbers. See the PARTS and scale smartly. ALL in ONE, MULTIPLY.

Speed

How Distance and Time Work Together

Every day, you deal with speed while walking to class, riding a bike, travelling by car, and even watching sports. Speed helps us describe how fast something moves and compare different types of motion.

Understanding speed is not just for maths exams. It is used in driving, sports timing, engineering, travel planning, and science. Once you learn the basic relationships between speed, distance, and time, you will start seeing them everywhere.

And the best part is that the maths behind speed is logical and very manageable once you understand the pattern.

Your Memory Anchor

Distance over time is speedy,
Less time means faster, nice and easy!
More distance wins the race,
Less time takes first place!"

Core Concept: What Is Speed?

Speed measures how quickly something travels over a distance.

It connects three key quantities:

- ✓ Speed → how fast something moves
- ✓ Distance → how far it travels
- ✓ Time → how long the journey takes

These three always form a relationship. If you know any two, you can find the third.

What These Questions Look Like

You might be asked:

- ✓ *Find the speed.*
- ✓ *Who is faster?*
- ✓ *How long will it take?*
- ✓ *Which arrives first?*
- ✓ *A journey is split into parts...*

If you understand distance, time, and speed, these become quick wins.

Speed Made Simple

You don't always need to calculate.

At the same speed:

- ✓ If time is halved, distance is halved.
- ✓ If time is doubled, distance is doubled.
- ✓ If distance is doubled, time is doubled.

Speed stays the same when distance and time change in the same ratio.

Worked Example

Question: A car travels 40 km in 1 hour. How far in 2 hours at the same speed?

Think smart:

Time doubles → distance doubles
40 → 80

✓ **Answer:** 80 km

Sometimes you can solve these using simple reasoning (no formula needed).

The One Formula You Must Know

Speed = Distance ÷ Time

From the rhyme: Distance over time is speedy.

Rearranging (same idea, different questions):

- ✓ **Distance** = Speed × Time
- ✓ **Time** = Distance ÷ Speed

You don't need three formulas, just one rule used smartly.

Worked Example

Question: A car travels at **80 km/h** for **2 hours**. How far does it go?

Calculate:

Step 1 — Values

Speed = 80 km/h
Time = 2 h

Step 2 — Formula

Distance = Speed × Time

Step 3 — Calculate

Distance = 80 × 2 = **160 km**

✓ **Answer:** 160 km

The FAST CHECK (5 seconds)

Before calculating anything, ask:

Question	Strategy	Reminder
Is the distance the same?	Compare time only **Less time = faster**.	Less time takes first place!
Is the time the same?	Compare distance only. **More distance = faster**.	More distance wins the race.
Are both different?	Use Speed = **Distance ÷ Time**	Distance over time is speedy.

Key Points to Remember

Units Matter

Always keep units consistent:

Distance	Time	Speed Unit
km	h	km/h
m	s	m/s

Mixing units leads to incorrect answers.

If speed is in km/h → use kilometres and hours

If speed is in m/s → use metres and seconds

Common Conversions

Original	Converted	Use For
30 minutes	0.5 hours	km/h
2 minutes	120 seconds	m/s
2 km	2000 m	m/s

Worked Example

Question: Speed = 60 km/h, time = 30 minutes. Find distance?

Convert time: 30 min → 0.5 hours (Because speed is in km/h, time must be in hours to keep units consistent.)

Calculate:

Distance = Speed × Time

Distance = 60 × 0.5 = 30

✓ **Answer:** 30 km

Journeys in Segments

Sometimes a journey is not one smooth trip. You may see:

- ✓ One part at one speed.
- ✓ Another part at a different speed.
- ✓ A rest or stop.
- ✓ Different distances in each part.

The Smart Method for Multi-Part Journeys

- ✓ Split the journey into segments.
- ✓ Use the correct formula for each segment
- ✓ Add results carefully.
- ✓ Only find averages at the end.

Never average speeds directly.

Finding Total Time (Most Common Case)

Worked Example

Question: A cyclist travels 10 km at 10 km/h, then 20 km at 20 km/h. How long does the journey take?

Think smart:

Formula: Time = Distance ÷ Speed

First part: 10 ÷ 10 = 1 hour

Second part: 20 ÷ 20 = 1 hour

✓ **Answer:** Total time = 1 + 1 = 2 hours

Question: A runner runs 6 km at 12 km/h, rests for 30 minutes, then runs 6 km at 6 km/h. How long does the journey take?

Think smart:

First run: 6 ÷ 12 = 0.5 h

Break: 30 min = 0.5 h

Second run: 6 ÷ 6 = 1 h

✓ **Answer:** Total time = 0.5 + 0.5 + 1 = 2 hours

Finding Total Distance

When time and speed are given for each part:

- ✓ Find distance for each segment.
- ✓ Add the distances.

Worked Example

Question: A car travels for 2 hours at 60 km/h, then 1 hour at 40 km/h. What is the total distance?

Think smart:

Formula: Distance = Speed × Time

First part: 60 × 2 = 120 km

Second part: 40 × 1 = 40 km

✓ **Answer:** Total distance = 120 + 40 = 160 km

Finding Average Speed for the Whole Journey

Average speed is found after the entire journey is combined.

- ✓ Find total distance.
- ✓ Find total time.
- ✓ Divide at the end.

$$\textbf{Average Speed} = \frac{\textbf{Total Distance}}{\textbf{Total Time}}$$

📌 **Not simply the average of speeds**

It means you **can't** just **add the speeds** and **divide by the number** of speeds when calculating average speed if the time or distance at each speed is different.

Worked Example

Question: A cyclist travels 10 km at 10 km/h and 20 km at 20 km/h. What is the average speed for the whole journey?

Think smart:

Total distance = 10 + 20 = 30 km

Total time = 1 + 1 = 2 h

Formula: Average speed = Total distance ÷ Total time
Average speed = 30 ÷ 2 = 15

✓ **Answer:** 15 km/h

Multiple-Choice Shortcut: "Who's Faster?"

When asked: Which is the fastest?

- ✓ Make units the same.
- ✓ If distance is equal → compare time.
- ✓ If time is equal → compare distance.
- ✓ Only then use distance ÷ time

The Most Common Traps

Avoid These Mistakes:

- ☒ Forgetting to convert minutes to hours.
- ☒ Mixing metres with kilometres.
- ☒ Averaging speeds directly.
- ☒ Ignoring rest or break time.
- ☒ Calculating when comparison was enough.

Remember: Match the units. Compare smart. Calculate only if needed.

Why This Chapter Matters

Speed appears in:

- ✓ Word problems.
- ✓ Comparisons.
- ✓ "Who arrives first" questions.
- ✓ Journey and travel problems.
- ✓ Multi-step reasoning

Mastering this chapter gives you easy marks under time pressure.

Try It Yourself

1. A car travels 150 km in 3 hours. What is its speed?
2. Two trains travel for the same time. Train A goes 200 km. Train B goes 240 km. Which is faster?
3. A runner moves at 10 m/s for 20 seconds. How far does she run?
4. A bike travels 12 km in 30 minutes. What is its speed in km/h?
5. A trip is split into two parts: 30 km at 60 km/h, then 30 km at 30
6. km/h. How long does the journey take?
7. A car drives 40 km, stops for 20 minutes, then drives 40 km at
8. the same speed. If its speed is 80 km/h, how long is the whole
9. journey?

Key Takeaway

Distance over time is speedy,
Less time means faster, nice and easy!
More distance wins the race,
Less time takes first place!

You don't need heavy formulas. You don't need long working.

Think smart. Eliminate fast. Solve with confidence.

SPEED CHEATSHEET: Think Fast, Don't Rush! (Selective + OC Tests)

MEMORY ANCHOR

Distance over time is speedy,
Less time means faster, nice and easy!
More distance wins the race,
Less time takes first place!

SPEED MADE SIMPLE (Constant Speed)

Time halved → **Distance halved**
Time doubled → **Distance doubled**
Distance doubled → **Time doubled**

} Speed stays same if D & T change in same ratio.

THE FAST CHECK (5-Second Strategy)

Before calculating, ask:

- Is **DISTANCE** same? → Compare **TIME** only. Less time = **FASTER**.
- Is **TIME** same? → Compare **DISTANCE** only. More distance = **FASTER**.
- Are **BOTH** different? → Use Formula: **S = D ÷ T**

THE ONE FORMULA

Speed = Distance ÷ Time
→ **Distance = Speed × Time**
→ **Time = Distance ÷ Speed**

- Just one rule, used smartly.

UNIT TRAPS & CONVERSIONS

Original		Convert To	Use For
30 minutes	→	0.5 hours	km/h
2 minutes	→	120 seconds	m/s
2 km	→	2000 m	m/s

KEY RULE: Match D & T units to Speed unit (km/h or m/s)!

JOURNEYS IN SEGMENTS

Split journey into **SEGMENTS** → Use formula for **EACH** segment → **ADD** results carefully (Total D, Total T)

AVERAGE SPEED

Average Speed = Total Distance ÷ Total Time

✗ **NEVER** average speeds directly!

COMMON TRAPS (Avoid!)

✗ Forgetting min → hour conversion
✗ Mixing metres with kilometres
✗ Averaging speeds directly
✗ Ignoring rest/break time
✗ Calculating when comparison is enough

Think smart. Eliminate fast. Solve with confidence.

Patterns & Sequences

Some numbers go up. Some go down. Some skip. Some switch. If you can see the pattern quickly, you will solve faster, avoid long working, eliminate wrong answers, and stay confident under time pressure. This chapter shows you how top selective students think.

Your Memory Anchor

In a sequence, spot the pattern

Increasing? Decreasing? What's happening?

Forward, backward, skip along

Find the rule and you can't go wrong!

What Is a Sequence?

A sequence is a list that follows a rule. It can be made of:

- **Numbers** → 2, 4, 6, 8...
- **Letters** → A, C, E, G...
- **Mixed** → 1A, 2C, 3E...

Your job is always the same: **Find the pattern. Apply the rule**.

The Three Pattern Types You Must Know

Type 1: Number Sequences

Numbers follow a mathematical rule. Common patterns include:

Pattern Type	Symbol	Description
Increasing	+	Adding a constant value.
Decreasing	−	Subtracting a constant value.
Multiplying	×	Each term multiplied by a factor.
Dividing	÷	Each term divided by a factor.
Skipping	⏭	Every second/third number.
Alternating	🔁	Two rules switching

Worked Example

Simple Increasing

Sequence: 3, 6, 9, 12, __

Analysis: Increasing by +3

✓ **Answer:** 15

From the anchor: "Increasing? Decreasing? What's happening?"

Type 2: Character (Letter) Sequences

Letters follow the alphabet. Remember the conversion:

A=1, B=2, C=3, ... Z=26

You can treat letters like numbers.

Worked Example

Letter Sequence

Sequence: B, D, F, H, __

As numbers: 2, 4, 6, 8 → +2

✓ **Answer:** J

From the anchor: "Forward, backward, skip along."

Type 3: Alternating & Mixed Sequences

Two patterns take turns, or numbers and letters change together.

Worked Example

Alternating Pattern

Sequence: 2, 5, 4, 7, 6, __

Pattern: +3, −1, +3, −1

✓ **Answer:** 9

Worked Example

Alternating Pattern

Sequence: 1A, 2C, 3E, 4G, __
Numbers: +1 → 5
Letters: +2 → I

✓ **Answer:** 5I

The FAST CHECK Method (5 Seconds)

Before doing any long working, ask:

- ✓ Is the sequence increasing or decreasing?
- ✓ Is it moving forward or backward?
- ✓ Is it skipping numbers or letters?
- ✓ Are two rules alternating?
- ✓ Are letters moving through the alphabet?

Number Sequences: The Smart Method

Step 1: Check the Differences

Subtract each term from the next.

Worked Example

Sequence: 5, 9, 13, 17, __

Differences: 9–5=**4**, 13–9=**4**, 17–13=**4**

Rule: +4

✓ **Answer: 21**

Step 2: Check for Multiply or Divide

If differences aren't equal, try × or ÷.

Worked Example

Sequence: 2, 6, 18, __

Pattern: ×3 each time

✓ **Answer:** 54

Step 3: Check for Skipping or Alternating

Look for jumps or two rules taking turns.

Worked Example

Sequence: 1, 4, 3, 6, 5, __

Pattern: +3, −1, +3, −1

✓ **Answer:** 8

Character Sequences: The Alphabet Trick

Convert letters into numbers, find the pattern, then convert back.

Worked Example

Sequence: C, F, I, L, __

As numbers: 3, 6, 9, 12

Rule: +3

Next: 15 → O

✓ **Answer:** O

Mixed Sequences: Numbers + Letters

Think in two tracks: number rule + letter rule.

Worked Example

Sequence: 2B, 4D, 6F, _

Numbers: +2 → 8

Letters: +2 → H

✓ **Answer:** 8H

Coding Patterns

Some questions don't look like sequences, they look like codes.

Digit Sum Coding

Example: 482 → 4 + 8 + 2 = 14

Thinking: "Small answer? Try digit sum first."

Digit Product Coding

Example: 231 → 2 × 3 × 1 = 6

Thinking: "Try digit product (multiply the digits) first."

Sum of Squares Coding

Example: 241 → $2^2 + 4^2 + 1^2$

= 4 + 16 + 1

= 21

Thinking: "*Digit sum is too small, product too big, so the squares are perfect*"

Outside Digits Only

Example: 735 → 7 × 5 = 35

Thinking: "*Maybe the middle digit is ignored.*"

Reverse and Combine

Example: 312 → 213

312 + 213 = 525

Thinking: *"If arithmetic fails, maybe digits are rearranged."*

Multiple-Choice Shortcut: Pattern Elimination

When stuck:

- ✓ Test only the first two steps of each option.
- ✓ Eliminate answers that break the rule.
- ✓ Only one option will follow the pattern exactly.

Advanced: Alternating Patterns

Sometimes the sequence uses two different rules that take turns. These are designed to confuse students who only look for one pattern.

Top scorers always check: Is something alternating?

Worked Example

Question: What comes next? 1, 4, 3, 8, 5, 12, ___

Analysis:

1 → 4 (+3)

4 → 3 (−1)

3 → 8 (+5)

8 → 5 (−3)

5 → 12 (+7)

Two patterns are taking turns:

Add odd numbers: +3, +5, +7

Subtract odd numbers: −1, −3, −5

Next step must follow the second rule: $12 - 5 = 7$

No guessing. No long working. Just structure.

Why This Chapter Matters

Patterns and sequences appear in:

- ✓ Number reasoning.
- ✓ Algebra foundations.
- ✓ Tables & graphs.
- ✓ Problem solving.
- ✓ Logic questions.
- ✓ Multiple-choice elimination.

This is one of the highest-value scoring skills in Selective & OC tests.

Try It Yourself

1. What comes next?
   3, 6, 11, 18, __
2. Find the next letter:
   A, D, G, J, __
3. What is the missing term?
   10, 20, 15, 25, 20, __
4. Complete the pattern:
   2B, 4D, 6F, __
5. Challenge:
   2, 7, 4, 11, 6, 15, __

Key Takeaway

In a sequence, spot the pattern

Increasing? Decreasing? What's happening?

Forward, backward, skip along

Find the rule and you can't go wrong!

You don't need to guess. You don't need long working.

Spot the pattern. Apply the rule. Solve with confidence.

Patterns are not magic. They're just organised thinking.

PATTERNS & SEQUENCES: CHEATSHEET & QUICK REFERENCE

MEMORY ANCHOR: In a sequence, SPOT THE PATTERN.
Increasing? ↗ Decreasing? ↘ What's happening? Forward →, backward ←, skip along ↺.
FIND THE RULE and you can't go wrong!

FOUNDATIONS & TYPES

WHAT IS A SEQUENCE?

[List (Numbers/Letters/Mixed)] follows a → [RULE] to make a → [PATTERN]

Goal: Find Pattern → Apply Rule.

THE 3 CORE TYPES

1. NUMBER SEQUENCES

- + (Add const.)
- − (Subtract const.)
- × (Multiply factor)
- ÷ (Divide factor)
- ⏭ (Skip steps)
- ⇄ (Alternating rules)

2. CHARACTER (LETTER) SEQUENCES

Follows Alphabet.
Treat as numbers.
A, B, C... → 1, 2, 3...
A=1, B=2, C=3 ... Z=26

3. ALTERNATING & MIXED

Two rules take turns
OR
Numbers + Letters change together.

THE SOLVING TOOLKIT

THE FAST CHECK METHOD (5 Secs)

- ✓ Increasing or Decreasing?
- ✓ Forward or Backward?
- ✓ Skipping?
- ✓ Alternating rules?
- ✓ Letters moving thru alphabet?

SMART METHODS (Step-by-Step)

Step 1: Check Differences (+/-)
↓
Step 2: Check Multiply/Divide (×/÷)
↓
Step 3: Check Skipping/Alternating (jumps or 2 rules)

CHARACTER TRICK

[Letter] — Convert A=1 → [Number Pattern] — Convert back → [Answer]

MIXED SEQUENCES

[Track 1:] Number Rule → [Combined Answer]
[Track 2:] Letter Rule → [Combined Answer]

SPECIAL CASES & SHORTCUTS

CODING PATTERNS (Non-sequence)

- Digit Sum (+ digits)
- Digit Product (× digits)
- Sum of Squares $(x^2 + y^2)$
- Outside Digits Only (ignore middle)
- Reverse & Combine (rearrange)

SHORTCUTS & ADVANCED

MCQ ELIMINATION FUNNEL

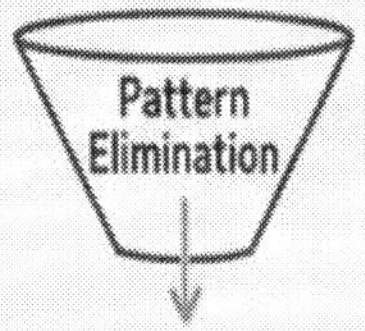

Test first two steps only
→ Eliminate rule-breakers
→ One option remains.

ADVANCED ALTERNATING

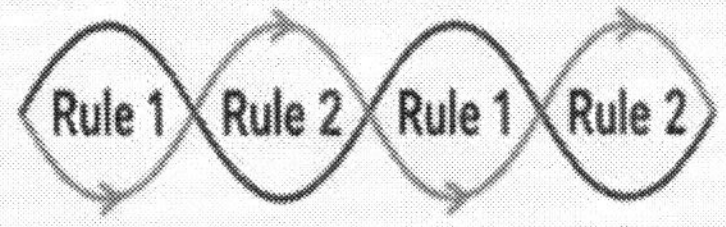

Watch for two different rules taking turns (e.g., +odd, then -even).

Patterns are organised thinking. No guessing. Just structure.

Operations & Mental Strategies

Every maths problem involves operations: adding, subtracting, multiplying, dividing, or combining them. Knowing what to do first and having mental shortcuts makes maths faster, easier, and far less stressful.

This chapter will help you:

- ✓ Understand the correct order of operations (BODMAS)
- ✓ Build strong mental maths strategies
- ✓ Solve problems confidently without always needing a calculator

Good operation skills do not just help in exams, they are useful for shopping, budgeting, cooking, travel, and everyday decision making.

Your Memory Anchor

Brackets first, Orders next,
Divide and Multiply left to right,
Add and Subtract to finish it right.

This rhythm helps you remember **BODMAS**:

- **B** — Brackets
- **O** — Orders (powers, roots)
- **D** — Division
- **M** — Multiplication
- **A** — Addition
- **S** — Subtraction

Division and multiplication share priority (left → right). Same with addition and subtraction.

Order of Operations

When a calculation has multiple operations, there is a correct order to follow. Without it, people would get different answers to the same problem.

Brackets → Orders (powers) → Division → Multiplication → Addition → Subtraction

Worked Example

$6 + 3 \times 4$

Steps:

If you add first → $9 \times 4 = \mathbf{36}$
If you multiply first → $6 + 12 = \mathbf{18}$

✓ **Answer: 18**, because multiplication comes before addition.

Real-Life Relevance

Order of operations appears everywhere:

- ✓ Calculating total cost with discounts and tax.
- ✓ Recipe adjustments in cooking.
- ✓ Travel budgeting and time calculations.
- ✓ Spreadsheet formulas.
- ✓ Construction measurements.

Mental maths strategies also help with:

- ✓ Estimating prices quickly.
- ✓ Checking calculator answers.
- ✓ Working efficiently in exams.

Order of Operations

Expressions with more than one operation.

Formula / Method

Follow this sequence:

1. Brackets first
2. Orders (powers/roots)
3. Divide and multiply left → right
4. Add and subtract left → right

Worked Example

$8 + 12 \div 3$

Steps:

Step 1 — Division first

$12 \div 3 = 4$

Step 2 — Addition

$8 + 4 = \mathbf{12}$

✓ Answer: 12

$(5 + 7) \times 2$

Steps:

Step 1 — Brackets first

$5 + 7 = 12$

Step 2 — Multiply

$12 \times 2 = \mathbf{24}$

✓ Answer: 24

Mental Maths Strategies

Calculations done efficiently in your head.

Formula / Method

Common strategies:

- ✓ Break apart numbers (partitioning).
- ✓ Round then adjust.
- ✓ Use known facts.
- ✓ Look for friendly numbers (10, 50, 100).

Breaking Apart

Sometimes numbers look difficult, but they are only difficult when we try to handle them all at once.

The Breaking Apart strategy, also called partitioning, works by splitting a number into smaller, friendlier pieces. Instead of calculating in one big step, you solve it in manageable chunks.

This method is especially useful for addition and multiplication, and it helps you see patterns in place value.

Worked Example

46 + 27

Steps:

Split 27: 46 + 20 + 7

66 + 7 = **73**

✓ Answer: 73

Round & Adjust

Some numbers are almost "friendly" numbers. They are just one or two away from something easier like 10, 50, 100, or 1000.

The Round & Adjust strategy makes a calculation easier by rounding a number first, solving the simpler problem, and then correcting the difference. This method is powerful for mental maths because round numbers are fast to work with.

Make it easy first, fix it after.

Worked Example

99 + 38

Steps:

Round 99 → 100: 100 + 38 = 138

Subtract extra 1: 138 – 1 = **137**

✓ **Answer:** 137

Nested Brackets

When brackets appear inside other brackets, calculations can look complicated. This is called nested brackets.

The key idea is simple. Always start with the innermost brackets and work your way outward step by step.

Trying to skip steps usually leads to mistakes. But if you peel the problem layer by layer, it becomes clear and manageable.

Solve the inside first, then work your way out.

Worked Example

3 (2 + (4 × 5))

Steps:

Multiply first inside: 4 × 5 = 20

Then: 2 + 20 = 22

Then: 3 × 22 = **66**

✓ **Answer:** 66

Estimation Strategy

Not every calculation needs an exact answer immediately. Sometimes you just need a close, reasonable value.

The Estimation Strategy helps you check your work, predict answers, and decide if a result makes sense. It often involves rounding numbers to friendly values before calculating.

Estimation builds number sense and prevents careless calculator errors.

If your exact answer is far from your estimate, something went wrong.

Sometimes an approximate answer is enough:

$$49 \times 19 \approx 50 \times 20 = 1000$$

Helpful for checking answers quickly.

Quick Checklist Summary

Step	**Question to Ask**	**Action**
1	Are there brackets?	Solve inside first
2	Any powers or roots?	Evaluate next
3	Multiply or divide?	Work left to right
4	Add or subtract?	Finish left to right
Mental Strategy	Can I simplify mentally?	Round, split, or estimate

Try It Yourself

Order of Operations

1. $6 + 5 \times 3$
2. $(10 - 4) \times 7$
3. $18 \div 3 + 9$
4. $4^2 + 6 \times 2$

Mental Maths

5. $48 + 29$
6. $99 + 56$
7. 25×4
8. Estimate: 51×18

Key Takeaway

Brackets first, Orders next,
Divide and Multiply left to right,
Add and Subtract to finish it right

You don't need to calculate everything.
Think smart. Eliminate fast. Solve with confidence.

OPERATIONS & MENTAL STRATEGIES: Faster, Easier, Less Stressful Maths.

Real-Life Relevance: Budgeting | Recipes | Data | Travel

THE ORDER OF OPERATIONS (BODMAS)

B ( ) [ ] { } — **BRACKETS**
Start Here.
Innermost first.

O x^2 $\sqrt{\ }$ — **ORDERS**
Powers & Roots.

D / M ÷ × ↔ — **DIVISION & MULTIPLICATION**
Share priority. Work Left → Right.

A / S + − ↔ — **ADDITION & SUBTRACTION**
Share priority. Finish Left → Right.

Memory Anchor:
Brackets first,
Orders next, Divide &
Multiply left to right,
Add & Subtract to
finish it right.

MENTAL MATHS TOOLKIT

(Think Smart, Calculate Fast)

BREAKING APART (Partitioning)
Split numbers into friendlier pieces.
Solve in manageable chunks.

ROUND & ADJUST.
Make it easy first (use 10, 50, 100).
Correct the difference after.

ESTIMATION STRATEGY.
Find a close, reasonable value quickly. Check if if exact answer makes sense.

[({ * })]

NESTED BRACKETS:
Peel layer by layer. Always start innermost and work outward.

QUICK PROCESS CHECKLIST

- ☑ 1. Are there **Brackets?** (Solve inside)
- ☑ 2. Any **Powers/Roots?** (Evaluate next)
- ☑ 3. **Multiply** or **Divide?** (Work Left → Right)
- ☑ 4. **Add** or **Subtract?** (Finish Left → Right)

Algebraic Thinking

Unknowns & Balance

Algebra is a **balance game, not a mystery.**

Some numbers are hidden.
Some numbers are waiting to be discovered.
Some numbers must keep both sides fair.

If you understand balance:

- You solve faster
- You avoid silly mistakes
- You spot exam traps quickly

This chapter helps turn algebra into a thinking skill, not just working out steps. And once that happens, algebra starts feeling logical and even enjoyable.

Your Memory Anchor

What I do to one side, I do to the other side too.
To find the unknown, I undo what was done to you.

This simple rhyme reminds us:

- ✓ Equations must stay balanced.
- ✓ We don't guess, we reason.
- ✓ We carefully undo operations.

What Is Algebraic Thinking?

Algebraic thinking means:

- ✓ Recognising patterns
- ✓ Using letters like x, y, or $n$for unknown numbers
- ✓ Understanding relationships between quantities
- ✓ Predicting outcomes logically

Real-Life Relevance

Algebra isn't just exam maths.
It's a way of **thinking about change, patterns, and relationships** in everyday life.

Whenever you estimate costs, plan travel, or track savings, you are already using algebraic thinking even if you do not write it with letters.

Algebra simply gives you a **clear language to describe patterns** so you can:

- ✓ Predict outcomes faster.
- ✓ Make smarter decisions.
- ✓ Solve unfamiliar problems confidently.

Let's see how it appears in real situations.

Shopping

If one notebook costs $3:

$$\textbf{Total cost} = \boldsymbol{3n}$$

(where $\boldsymbol{n}$ = number of notebooks)

Instead of calculating again each time, you have a rule.

Travel

Distance depends on speed and time:

$$\boldsymbol{d = st}$$

This helps you:

- ✓ Estimate travel time.
- ✓ Plan trips.
- ✓ Compare routes quickly.

Saving Money

Saving regularly creates predictable patterns.

Example: Saving $10 weekly means your total savings grow by $10 each week. If **n** represents the number of weeks, you can easily predict how much you will have later.

Writing Algebraic Expressions

We translate words or situations into algebra. This is essential for exam word problems.

Method: Pattern Representation

Use a variable such as $\boldsymbol{x}$.

Words	**Algebra**
Five more than a number	$x+5$
Twice a number	$2x$
Seven less than a number	$x-7$

Worked Example

Write an expression for: Four times a number plus six. Let the number = $\boldsymbol{x}$

Expression: $\boldsymbol{4x+6}$

Solving Equations (Balance Method)

What are we solving?

- ✓ We want the value of $\boldsymbol{x}$.
- ✓ To see it clearly, we get $\boldsymbol{x}$ by itself while keeping balance.

Balance Rules

If you see	**Do this to BOTH sides**
+ number	Subtract it
− number	Add it
× number	Divide by it
÷ number	Multiply by it

Always keep both sides fair.

Important

The equation is already balanced.
We're simply revealing the hidden number.

Worked Example

$\boldsymbol{x + 8 = 20}$

Subtract 8 from both sides:

$\boldsymbol{x + 8 - 8}$	$=$	$\boldsymbol{20 - 8}$
x	$=$	12

Check: $12 + 8 = \mathbf{20}$

✓ **Answer:** 20

Worked Example

$\boldsymbol{3x = 21}$

Divide both sides by 3: $x = 7$

Check: $3 \times 7 = 21$

Worked Example

$2x + 5 = 17$

Divide by 2: $x = 6$

Check: $2(6) + 5 = 17$

Common Trap to Avoid

Question: Two numbers add to 20. One number is x. The other is **4 more than** x. Find both numbers.

Steps:

Step 1 — Translate carefully

Second number: $x + 4$

Total: $x + (x + 4) = 20$

Brackets show the whole second number.

Step 2 — Simplify

$2x + 4 = 20$

Step 3 — Solve

Subtract 4: $2x = 16$

Divide by 2: $x = 8$

Step 4 — Find both numbers

First number: $x = 8$

Second number: $x + 4 = 12$

✓ **Answer:** 8 and 12

Why Students Lose Marks

Common mistakes:

- ☒ Forgetting both numbers.
- ☒ Missing brackets.
- ☒ Only solving for one value.
- ☒ Not checking the total.

Always reread the question.

Quick Checklist Summary

Before finishing:

- ✓ Did I translate words correctly?
- ✓ Did I keep both sides balanced?
- ✓ Did I isolate the variable logically?
- ✓ Did I answer all parts?
- ✓ Did I check my answer?

Try It Yourself

1. Find x: **x + 7 = 18**
2. Solve: **4x = 32**
3. What goes in the box? □ **– 9 = 11**
4. A function machine: **×3**, then **−2**. What is the output if the input is **5**?
5. Trial and improvement: A number is doubled and then **6** is added to get **20**. What is the number?
6. Logic puzzle: Two numbers total **30**. One is **x**. The other is **3** more than **x**. Find **x**.

   Multiple choice: Which value of x makes this true? **2x + 1 = 9**

   A) 3 B) 4 C) 5 D) 6

Key Takeaway

What I do to one side, I do to the other side too.
To find the unknown, I undo what was done to you.

You don't need complicated methods.
Undo. Balance. Check. Solve with confidence.

ALGEBRAIC THINKING: UNKNOWNS & BALANCE

It's a balance game, not a mystery. Logic over guessing.

MEMORY ANCHOR

What I do to one side, I do to the other side too.
To find the unknown, I undo what was done to you.

WHAT IS IT?

- ☑ Recognising patterns
- ☑ Using variables (x, y, n) for unknowns
- ☑ Understanding relationships
- ☑ Predicting outcomes logically

REAL-LIFE RELEVANCE

Shopping: Total cost rules (e.g., 3n$ for n items at $3 each).

Travel: Distance = Speed × Time ($d = st$). Estimate & plan.

Saving: Predictable growth patterns. Track progress.

THE BALANCE METHOD (Goal: Isolate x)

IF YOU SEE:	DO THIS:	IF YOU SEE:	DO THIS:
+	− (Subtract)	−	+ (Add)
×	÷ (Divide)	÷	× (Multiply)

<u>MUST</u> do the same to <u>BOTH</u> sides to keep it fair!

TRANSLATING WORDS TO MATH

Words / Situation → Algebraic Expression

- More than / Sum → +
- Less than / Difference → −
- Times / Product of → ×
- Divided by / Quotient → ÷

QUICK CHECKLIST & TRAPS

- ☑ Translate carefully?
- ☑ Maintain balance?
- ☑ Isolate logically?
- ☑ Check answer!

AVOID:

- ✗ Missing brackets (e.g., $2(x+4)$)
- ✗ Not answering all parts
- ✗ Forgetting both numbers

KEY TAKEAWAY: Undo. Balance. Check. Solve with confidence.

Measurement & Units

Measure Like a Pro (Convert Fast. Read Smart. Avoid Traps.)

Length, mass, and capacity appear everywhere in Selective and OC tests.

If you master units and scales, you will convert in seconds, avoid careless mistakes, eliminate wrong answers quickly, and estimate like a top student.

This chapter turns measurement into a scoring weapon.

When you master measurement:

- ✓ You convert quickly and accurately.
- ✓ You avoid careless mistakes.
- ✓ You eliminate wrong answers fast.
- ✓ You estimate like a confident mathematician.

This chapter will help turn measurement into a **scoring advantage**, not a guessing game.

Your Memory Anchor

Small to big? Divide.
Big to small? Multiply.
Different units? Don't combine.
Convert first, then solve every time.

The Three Measurements You Must Know

Length (How long?)

Used for distance, height, width.

- ✓ millimetres (mm)
- ✓ centimetres (cm)
- ✓ metres (m)
- ✓ kilometres (km)

Units:	mm	→	cm	→	m	→	km

Object	Best Unit
Paper thickness	mm
Pencil length	cm
Classroom length	m
Distance between cities	km

Mass (How heavy?)

Used for weight of objects.

- ✓ milligrams (mg)
- ✓ grams (g)
- ✓ kilograms (kg)
- ✓ tonnes (t)

Units:	g	→	kg	→	t

Object	Best Unit
Apple	g
Backpack	kg
Truck Load	t (tonne)

Capacity (How much it holds?)

Used for liquids.

- ✓ millilitres (mL)
- ✓ litres (L)

Units:	mL	→	L

Object	Best Unit
Medicine dose	mL
Water bottle	L

Real-Life Relevance

Measurement skills aren't just test skills:

- ✓ Reading medicine doses safely
- ✓ Comparing product prices
- ✓ Following recipes correctly
- ✓ Estimating travel distance or time
- ✓ Understanding sports performance stats

Strong measurement sense makes everyday maths easier.

The Unit Ladder Method

This is the most intuitive method. Changing units while keeping the same quantity.

Example:

3 m = 300 cm
Same length, different unit.

Formula

Big → Small = Multiply
Small → Big = Divide

Common Metric Steps

Length

From	To	Direction	Operation
km → m	Big → Small	Multiply	× 1000
m → cm	Big → Small	Multiply	× 100
cm → mm	Big → Small	Multiply	× 10
mm → cm	Small → Big	Divide	÷ 10
cm → m	Small → Big	Divide	÷ 100
m → km	Small → Big	Divide	÷ 1000

Mass

From	To	Direction	Operation
kg → g	Big → Small	Multiply	× 1000
g → mg	Big → Small	Multiply	× 1000
mg → g	Small → Big	Divide	÷ 1000
g → kg	Small → Big	Divide	÷ 1000

Capacity

From	To	Direction	Operation
L → mL	Big → Small	Multiply	× 1000
mL → L	Small → Big	Divide	÷ 1000

Pattern You Should Notice

When the unit gets **smaller**, the number must get **bigger** → multiply.
When the unit gets **larger**, the number must get **smaller** → divide.

That's the logic behind your Memory Anchor:

Small to big? Divide.
Big to small? Multiply.

When students understand why the number changes direction, conversions become automatic instead of mechanical.

Worked Example

Convert **4.5 kg to grams.**

Steps:

Step 1 — Direction

kg → g (big to small) → multiply.

Step 2 — Convert

4.5 × 1000 = 4500 g.

✓ **Answer:** 4500 g

Worked Example

Convert 750 cm to metres.

Steps:

Step 1 — Direction

cm → m (small to big) → divide.

Step 2 — Convert

750 ÷ 100 = 7.5 m.

✓ **Answer:** 7.5 m

Unit Conversions: The No-Fail Rule

When converting between metric units, each step involves multiplying or dividing by powers of 10.

- ✓ Moving from **smaller** units to **larger** units requires **division**.
- ✓ Moving from **larger** units to **smaller** units requires **multiplication**.

Because metric units are based entirely on powers of 10, conversions can always be done quickly and safely.

Metric Power Table (No-Fail Toolkit)

kilo	hecto	deca	base	deci	centi	Milli
k	h	da	—	d	c	m
3	2	1	0	-1	-2	-3

Universal Conversion Shortcut

Move = (Target Power) – (Start Power)

- Move is **positive** → decimal moves left (**divide**)
- Move is **negative** → decimal moves right (**multiply**)

Worked Example

Question: Convert 250 **cm** into **metres**.

Think smart:

$$\text{cm} = -2$$
$$\text{m} = 0$$

Move:

$$= (\text{Target Power}) - (\text{Start Power})$$
$$= 0 - (-2)$$
$$= 2$$

Move decimal left 2 places: 250 → 2.5

✓ **Answer:** 250 cm = 2.5 m

Worked Example

Question: Convert 3.6 **m** into **centimetres**.

Think smart:

$$\text{cm} = 0$$
$$\text{m} = -2$$

Move:

$$= (\text{Target Power}) - (\text{Start Power})$$
$$= (-2) - 0$$
$$= -2$$

Move decimal right 2 places: 3.6 → 360

✓ **Answer:** 3.6 m = 360 cm

Worked Example

Question: Convert 4.5 **m** into **kilometres**.

Think smart:

m = 0
km = 3

Move:

= (Target Power) − (Start Power)
= 3 − 0
= 3

Move decimal left 3 places: 4.5 → 0.0045

✓ **Answer:** 4.5 m = 0.0045 km

Worked Example

Question: Convert 350 **cm** into **deca-metres** (dam).

Think smart:

cm = −2
dam = +1

Move:

= (Target Power) − (Start Power)
= 1 − (−2)
= 1 + 2
= 3

Move decimal left 3 places: 350 → 0.35

✓ **Answer:** 350 cm = 0.35 dam

What These Questions Look Like

You might see:

- ✓ How many millilitres are in 3.5 L?
- ✓ A bottle holds 750 mL. How many bottles fill 3 L?
- ✓ A jug holds 1.2 L. How much more is needed to make 2 L?
- ✓ Which container holds the greatest volume?
- ✓ Which unit is most suitable for measuring this object?

These are about understanding size, not doing hard arithmetic.

FAST CHECK (5 Seconds Before You Calculate)

Before working, ask:

- ✓ Are the units the same?
- ✓ Do I need to convert first?
- ✓ Does the answer seem reasonable (too big / too small)?

Most mistakes happen when students skip this step.

Worked Example

Basic Conversion

Question: How many **millilitres** are in **3.5 L**?

Think smart:

$$1\text{ L} = 1000\text{ mL}$$
$$3.5 \times 1000 = 3500$$

✓ **Answer:** 3500 mL

Finding the Difference

Question: A jug holds **1.2 L**. How much more is needed to make **2 L**?

$$\mathbf{2 - 1.2 = 0.8}$$

✓ **Answer:** 0.8 L

Division with Conversion

Question: A bottle holds **750 mL**. How many bottles make **3 L**?

Convert:

$3 \text{ L} = 3000 \text{ mL}$
$3000 \div 750 = 4$

✓ **Answer:** 4 bottles

Worked Example

Subtraction with Conversion

Question: A container has **2.4 L. 900 mL** is poured out. How much is left?

Convert:

$2.4 \text{ L} = 2400 \text{ mL}$
$2400 - 900 = 1500 \text{ mL}$

✓ **Answer:** $1500 \text{ mL} = 1.5 \text{ L}$

Combining Measurements

You cannot add or subtract measurements written in different units directly.

Why?

Because you can only add or subtract measurements when they are in the same unit.

Metres and centimetres are different sizes, so they must be converted first.

It is like trying to add:

- ☒ 2 dollars + 40 cents
- ☒ 2 apples + 40 oranges

2 m + 40 cm = 42 m (This is also wrong because 40 cm is not 40 metres; it is only part of a metre)

You must convert to the same unit before combining.

Correct Method

Step 1: Convert to the same unit

→ Convert metres into centimetres:

→ 2 m = 200 cm

Step 2: Add using the same unit

→ 200 cm + 40 cm = 240 cm

Step 3: Convert back if needed

→ 240 cm = 2.4 m

✓**Answer:** 2 m + 40 cm = 240 cm = 2.4 m

Choosing the Right Unit

Which unit is most appropriate?

Measurement	Best Unit
Mass of an apple	g
Mass of a person	kg
Water in a cup	mL
Fuel in a car	L
Distance between cities	km

Quick Check

If a unit looks too tiny or too huge, it's probably wrong.

Worked Example

Selecting Units

Question: Which **unit** is best to **measure** the **mass** of a watermelon?

A) milligrams B) grams C) kilograms D) tonnes

Think smart:

Too small → milligrams (mg)
Used for very **tiny masses**, like a grain of salt or medicine powder.

Small → grams (g)
Used for **light objects**, like a paperclip, a chocolate bar, or an apple.

Just right → kilograms (kg)
Used for **heavier items**, like a watermelon, a bag of rice, or a dog.

Too big → tonnes (t)
Used for **extremely heavy things**, like a car, an elephant, or a truckload of sand.

A watermelon weighs a few kilograms

✓ **Answer:** C) kilograms

Comparing

Question: Which is heavier: **850 kg** or **0.9 t**?

Think smart:

To compare masses, we must convert them into the same unit.

0.9 t = 900 kg

Now compare:

900 > 850

✓ **Answer:** 0.9 t is heavier

Multiple-Choice Shortcuts

Unit Elimination

- ✓ Cross out answers in the wrong unit
- ✓ Eliminate values that are clearly too big or too small
- ✓ Convert only what remains

Smart Comparison

To compare two measurements:

- ✓ Convert only one, or
- ✓ Round to friendly numbers

Estimation: Think Like a Top Student

Estimation checks reasonableness, not exactness.

Object	Approximate Value
Paperclip	≈ 1 g
Textbook	≈ 500 g – 1 kg
Water bottle	≈ 500 mL – 1 L
Classroom length	≈ 8–10 m

Why This Chapter Matters

Measurement appears in:

- ✓ Volume and capacity problems
- ✓ Unit conversion questions
- ✓ Real-life word problems
- ✓ Comparison and estimation tasks

Try It Yourself

1. Convert 4.2 L into millilitres.
2. A bottle holds 500 mL. How many bottles make 2.5 L?
3. A container has 3 L of water. 750 mL is used. How much is left?
4. Which is larger: 1.8 kg or 1750 g?
5. A ruler shows marks every 2 cm. The pointer is 3 marks after 10 cm. What is the reading?
6. Which unit is best to measure the length of a classroom?
   A) mm B) cm C) m D) km

Key Takeaway

Small to big? Divide.
Big to small? Multiply.
Different units? Don't combine.
Convert first, then solve every time.

Measurement problems aren't hard.
They just need clear thinking, careful units, and smart estimation.

MEASUREMENT & UNITS: CHEATSHEET (Convert Fast. Read Smart.)

THE THREE MEASUREMENTS

Length (Distance)

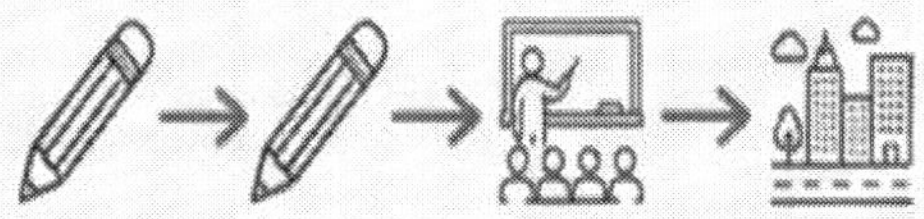

mm (pencil tip) → cm (pencil) → m (classroom) → km (city distance)

Mass (Weight)

g (apple) → kg (backpack) → t (truck load)

Capacity (Liquid)

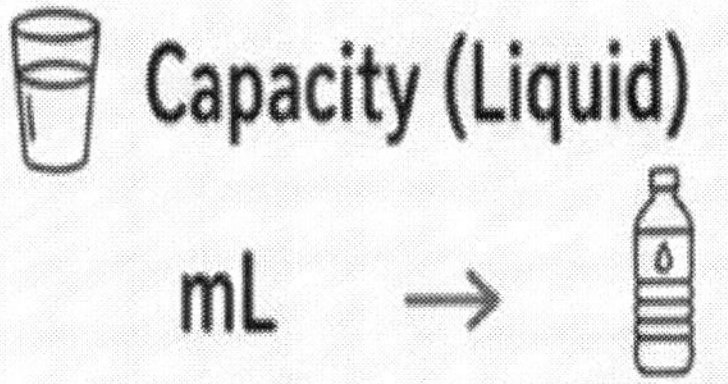

mL (medicine dose) → (water bottle)

MEMORY ANCHOR

Small → Big? DIVIDE (÷)

Big → Small? MULTIPLY (×)

Different Units? DON'T COMBINE.

Convert first, then solve.

CONVERSION ENGINE (Powers of 10)

Metric Power Table

k (kilo)	h (hecto)	da (deca)	BASE UNIT	d (deci)	c (centi)	m (milli)
3	2	1	0	-1	-2	-3

Universal Shortcut

Move = (Target Power) – (Start Power)

← Positive Move (+): Decimal LEFT (Divide)

Negative Move (-) →: Decimal RIGHT (Multiply)

QUICK TOOLKIT & CHECKS

Standard Equivalences

- 1 cm = 10 mm
- 1 m = 100 cm
- 1 km = 1000 m
- 1 kg = 1000 g
- 1 t = 1000 kg
- 1 L = 1000 mL

FAST CHECK ROUTINE (5 Seconds)

- ☑ Are units the same?
- ☑ Do I need to convert?
- ☑ Is answer reasonable? (Estimation)

Combining

RULE: Never add/subtract different units. Convert to matching unit first.

Time, Timetables & Calendars

Master the Clock and the Calendar. Learn It Once, Use It Everywhere.

Time questions aren't hard... they're tricky. Most students lose marks because they rush conversions, forget to carry hours and minutes, misread timetables, or count days one by one. If you can control time and dates, you will avoid careless mistakes, solve faster than most students, and stay calm in multi-step problems.

Your Memory Anchor

12-hour has am/pm.
24-hour has no am/pm.
Break time into steps.
Follow the timetable line.
Weeks repeat every 7.
Dates move in chunks.

Time Formats

Understanding the two time formats is essential for conversion questions and avoiding common traps.

12-Hour Time (am / pm)

- ✓ Uses **am** (midnight → before midday)
- ✓ Uses **pm** (midday → before midnight)

Examples: 7:30 am, 12:45 pm, 9:05 pm

Things to Remember

12:00 am = **midnight**.
12:00 pm = **midday**.

24-Hour Time (No am/pm)

Runs from **00:00** to **23:59**

After 12:59 → keep adding hours

12-Hour	**24-Hour**
2:15 pm	14:15
6:40 pm	18:40
12:00 am	00:00

12-Hour ↔ 24-Hour Conversion (FAST Method)

Converting pm → 24-hour

Add 12 (except 12 pm stays as 12):

3:20 pm → 15:20

11:45 pm → 23:45

Converting am → 24-hour

Same number (except midnight):

6:10 am → 06:10

9:35 am → 09:35

Converting 24-hour → 12-hour

If hour ≥ 13, subtract 12 and add pm

18:30 becomes 18 – 12 = 6:30 pm

If hour < 12, keep the hour the same and add am

09:20 becomes 9:20 am

Time Conversion Table

24-Hour Time	12-Hour Time
00:00	12:00 AM (Midnight)
01:00	1:00 AM
02:00	2:00 AM
03:00	3:00 AM
04:00	4:00 AM
05:00	5:00 AM
06:00	6:00 AM
07:00	7:00 AM
08:00	8:00 AM
09:00	9:00 AM
10:00	10:00 AM
11:00	11:00 AM
12:00	12:00 PM (Noon)
13:00	1:00 PM
14:00	2:00 PM
15:00	3:00 PM
16:00	4:00 PM
17:00	5:00 PM
18:00	6:00 PM
19:00	7:00 PM
20:00	8:00 PM
21:00	9:00 PM
22:00	10:00 PM
23:00	11:00 PM

Finding the Duration (Finish – Start)

This is one of the most common OC and Selective exam skills.

You are given:

- A start time
- A finish time

You must find:

How long something lasted.

⚠ **The Golden Rule**

Duration = Finish Time – Start Time

If AM/PM appears, convert both times to 24-hour time first.

This prevents mistakes when crossing midday or midnight.

Method: Balance (Step Forward)

Instead of jumping straight to subtraction:

1. Move from the start time to the next full hour
2. Add full hours
3. Add remaining minutes
4. Combine everything

This avoids borrowing errors.

Worked Example

Question: A lesson starts at **9:35 am** and ends at **11:20 am**. How long is the lesson?

Steps:

Step 1: Move to the next hour

9:35 → 10:00 = 25 minutes

Step 2: Add full hours

10:00 → 11:00 = 1 hour

Step 3: Add remaining minutes

11:00 → 11:20 = 20 minutes

Step 4: Add everything

= 1 hour + (25 + 20) minutes

= 1 hour + 45 minutes

= **1 hour 45 minutes**

✓ **Answer:** 1 hour 45 minutes

Worked Example

Question: A concert begins at **11:50 am** and finishes at **1:15 pm**. How long is it?

Steps:

Step 1: Convert to 24-hour

11:50 am = 11:50

1:15 pm = 13:15

Step 2: Step forward (Balance)

11:50 → 12:00 = 10 min

12:00 → 13:00 = 1 hour

13:00 → 13:15 = 15 min

Total = 1 hour + (10 + 15) min

= **1 hour 25 minutes**

✓ **Answer:** 1 hour 25 minutes

Finding the Finish Time (Start + Duration)

👉 The Golden Rule

Elapsed Time = Start → Change → End

Move forward in steps. Don't jump in one go.

FAST Method for Elapsed Time

1. **F** – First, go to the next hour (Use **F only** if the minutes will pass 60).
2. **A** – Add full hours.
3. **S** – Subtract what you used.
4. **T** – Time left goes on at the end.

Worked Example

Question: A movie starts at 3:45 pm and lasts 1 hour 35 minutes. What time does it finish?

Steps:

Check minutes: 45 + 35 = 80 → passes 60 → **F is needed.**

F – First, go to the next hour

3:45 → 4:00 = 15 minutes

A – Add full hours

4:00 → 5:00 = 1 hour

S – Subtract what you used

35 minutes – 15 minutes = 20 minutes left

T – Time left goes on at the end

5:00 + 20 minutes = 5:20 pm

✓ **Answer:** 5:20 pm

Worked Example

Question: A bus leaves at 08:40. It travels for 55 minutes, stops for 15 minutes, then continues for 1 hour 20 minutes. What time does it arrive?

Step 1: First travel (**55 minutes**)

Check minutes: 40 + 55 = 95 → passes 60 → **F is needed**

F – First, go to the next hour

08:40 → **09:00** = 20 min

A – Add full hours

None (0 hours)

S – Subtract what you used

55 – 20 = **35 min**

T – Time left goes on at the end

09:00 + 35 min = **09:35**

Step 2: Stop (**15 minutes**)

Check minutes: 35 + 15 = 50 → does not pass 60 → **F not needed**

09:35 + 15 min = **09:50**

Step 3: Second travel (1 hour 20 minutes)

Check minutes:

50 + 20 = 70 → passes 60 → **F is needed**

F – First, go to the next hour
09:50 → 10:00 = 10 min

A – Add full hours
10:00 → **11:00** = 1 hour

S – Subtract what you used
20 – 10 = **10 min**

T – Time left goes on at the end
11:00 + 10 min = 11:10

✓ **Answer:** 11:10

Note

You do not need FAST method every time.

If the hours and minutes look easy to add, you can just add them directly.

Use FAST when the minutes are tricky or cross into the next hour.

Time Questions (Quick Recognition Guide)

Not all elapsed-time questions are the same.

If the question asks...	Use this method
How long did it take?	Finish – Start
What time does it finish?	Start + Duration
What time did it start?	Finish – Duration

Recognising the question type quickly prevents careless mistakes.

Adding Large Numbers of Hours

Big "hours later" questions are designed to make students count one by one. But clocks repeat, so it's better to use cycles instead.

The Golden Rule

Divide by **12** (for am/pm) or divide by **24** (for full days).

Ignore full cycles. Use the remainder.

Method 1: "Mod 12" (12-Hour Clock)

1. **Add** the hours.
2. **Divide** by 12 → keep the remainder.
3. Count am/pm **flips** (every 12 hours = one flip).

Quick AM/PM Flip Rule

- ✓ **Odd** number of 12s → **flip** AM/PM (Flip AM to PM **or** PM to AM)
- ✓ **Even** number of 12s → stays the **same** (AM or PM stays the same)

If remainder = 0, the hour is 12, not 0

Mental Shortcuts

+24	Same time, next day.
+12	Same hour, flip am/pm.
+48	Same time, two days later.

Worked Example

Question: 2:00 pm + 53 hours =?

Steps:

Step	**Calculation**	**Result**
Add hours	2 + 53 = 55	
Divide by 12	55 ÷ 12 = 4 remainder 7	Hour = 7
Count flips	4 flips → even	Stays pm

✓ **Answer:** 7:00 pm

Method 2: "Mod 24" (24-Hour Clock) — Even Faster!

1. **Convert** to 24-hour
2. **Add** hours
3. **Divide** by 24 → keep the remainder
4. **Convert** back

Worked Example

Question: 2:00 pm + 53 hours =?

Steps:

Step	Calculation	Result
Convert to 24-hour	2:00 pm → 14:00	
Add hours	14:00 + 53 = 67	
Divide by 24	67 ÷ 24 = 2 remainder 19	19:00
Convert back	19:00 → 7:00 pm	7:00 pm

✓ **Answer:** 7:00 pm

Time Zones (Basic)

Time Zone Rule (Easy Way to Remember)

- ✓ When you **move east**, the time gets **later**.
- ✓ When you **move west**, the time gets **earlier**.

Why?

The Earth spins from west to east, so places in the east see the sun first. That's why they are ahead in time.

Example:

- Sydney is ahead of Perth → Sydney is further **east**, so it sees the sun earlier and has a **later time**.
- Perth is behind Sydney → Perth is further **west**, so it sees the sun later and has **an earlier time**.

Worked Example

Question: Sydney is in the **east**, and the time there is **6:00 pm**. The time difference between Sydney and Perth is **3 hours**. What is the time in Perth?

Think smart:

1. Sydney is further east, so its time is ahead.
   "Ahead" means Sydney's clock shows a later time than Perth's.
2. Perth is to the west, so its time is earlier.
   "Earlier" means Perth's clock shows an earlier time than Sydney's.
3. So to find Perth's time (the earlier time), subtract the difference from Sydney's time

Perth is 3 hours behind Sydney.**6:00 pm – 3 hours = 3:00 pm**

✓ **Answer:** 3:00 pm

Calendar Questions

Calendar questions follow patterns, not counting. Control the days without thinking like a robot.

The Week Wheel (Your Secret Weapon)

Picture a circle with the 7 days of the week:

Monday	Tuesday	Wednesday	Thursday	Friday	Saturday	Sunday	back to Monday

What the Wheel Teaches You

- ✓ Every 7 days, you land on the same day (one full circle).
- ✓ So, **ignore** full weeks.
- ✓ Only the **leftover** days (remainder) matter.
- ✓ Count **forward** from today.

Worked Example

Question: Today is Tuesday. What day will it be in 45 days?

Steps:

Step 1: Divide by **7**

45 ÷ 7 = 6 full weeks **remainder 3**

Step 2: Ignore the full weeks

6 full circles bring you back to **Tuesday**

Step 3: Count forward the remainder

Move forward **3 days from Tuesday**:

Wednesday (1)

Thursday (2)

Friday (3)

✓ **Answer:** Friday.

Finding the Date (Big Numbers = Use Chunks)

👉 **The Golden Rule**

Break into months first, then use leftovers.

Never count day by day.

FAST METHOD for "Future Date"

1. **Subtract** to the end of the current month.
2. **Jump** whole months.
3. Use the **remaining** days.

Month Days Cheat Table

Jan	Feb	Mar	Apr	May	Jun	Jul	Aug	Sep	Oct	Nov	Dec
31	28*	31	30	31	30	31	31	30	31	30	31

**February has 29 days in a leap year.*

Question: If 1 January is Monday, what day is 15 January?

Think smart:

14 days later → exactly 2 weeks → same day

✓ **Answer:** Monday

Question: If today is 10 March, what will be the date 95 days later?

Steps:

Step 1: Finish the Current Month

March has 31 days.

From 10 March → 31 March = 21 days

Remaining:

95 – 21 = 74 days

Step 2: Jump Whole Months

April = 30 days

→ 74 – 30 = 44

May = 31 days

→ 44 – 31 = 13

So we are now in June, with 13 days remaining.

Step 3: Add Remaining Days

13 days into June

→ 13 June

✓ **Answer:** 13 June

Common Traps

- ☒ Mixing am and pm.
- ☒ Forgetting 60 minutes = 1 hour.
- ☒ Adding everything at once.
- ☒ Counting days one by one.
- ☒ Forgetting February and leap years.

Why This Chapter Matters

- ✓ These skills appear in:
- ✓ Elapsed time.
- ✓ Timetables.
- ✓ Time zones.
- ✓ Multi-step word problems.
- ✓ Calendar reasoning.

Try It Yourself

1. Convert to 24-hour time:
   a) 7:45 pm b) 12:00 am c) 9:05 am
2. A class starts at 10:25 am and lasts 1 hr 50 min. When does it end?
3. 2:00 pm + 53 hours =?
4. When it is 4:00 pm in Sydney, it is 2 hours earlier in Adelaide. What time is it there?
5. Today is Thursday. What day will it be in 23 days?
6. What date is 18 days after 12 June?
7. In a leap year, what is 3 days after 27 February?
8. If 5 August is Monday, what day is 26 August?

Key Takeaway

12-hour has am/pm.

24-hour has no am/pm.

Break time into steps.

Use cycles for big numbers.

Use the Week Wheel for days.

Jump by months for dates.

You don't need to rush.

Think in stages. Move in chunks. Control the clock and the calendar.

MASTERING TIME & CALENDARS: CHEATSHEET

Learn It Once, Use It Everywhere. Think in Stages, Control the Clock.

TIME FORMATS & CONVERSIONS

12-Hour (am/pm) vs. 24-Hour

- 12-Hour: Uses am (midnight → before midday), pm (midday → before midnight). 12:00 am = midnight, 12:00 pm = midday.
- 24-Hour: 00:00 to 23:59. No am/pm. After 12:59 → keep adding hours.

2:15 pm	↔	14:15
6:40 pm	↔	18:40
12:00 am	↔	00:00

Conversion Rules (FAST Method)

- pm → 24h: Add 12 (except 12 pm stays 12). e.g., 3:20 pm → 15:20
- am → 24h: Same number (except midnight 12 → 00). e.g., 6:10 am → 06:10
- 24h → 12h: If hour ≥ 13, subtract 12, add pm. If < 12, keep hour, add am. e.g., 18:30 → 6:30 pm

ELAPSED TIME & LARGE JUMPS

Elapsed Time (FAST Method)

Golden Rule: Start → Change → End. Move in steps.

- F – First, go to next hour (if mins > 60).
- A – Add full hours.
- S – Subtract what you used.
- T – Time left goes at end.

Start Time —(F)→ Next Hour —(A)→ Hours Later —(T)→ End Time

Adding Large Hours (Cycles)

Golden Rule: Divide by 12 (am/pm) or 24 (full days). Use remainder.

Mod 12 (12-Hour Clock):

1. Add hours.
2. Divide by 12, keep remainder.
3. Count am/pm flips (Odd=Flip, Even=Same).

Mod 24 (24-Hour Clock):

1. Convert to 24h.
2. Add hours.
3. Divide by 24, keep remainder.
4. Convert back.

Shortcuts:

- +24h = Same time, next day.
- +12h = Same hour, flip am/pm.
- +48h = Same time, two days later.

TIME ZONES & CALENDARS

Time Zones (Basic Rule)

WEST (Earlier, -) ←— TIME —→ EAST (Later, +)

Move East → Time gets LATER.
Move West → Time gets EARLIER.

Calendar Patterns (Week Wheel)

The Week Wheel: Mon → Tue → Wed → Thu → Fri → Sun → Mon

Ignore full weeks (÷7). Count remainder forward from today.

Finding Dates (Chunks)

Golden Rule: Break into months, use leftovers. Never count days.

FAST METHOD for Future Date:

1. Finish current month.
2. Jump whole months.
3. Add remaining days.

Month Days Cheat Table		
Jan(31)	Feb(28/29*)	Mar(31)
Apr(30)	May(31)	Jun(30)
Jul(31)	Aug(31)	Sep(30)
Oct(31)	Nov(30)	Dec(31)

*Leap year.

Common Traps: Mixing am/pm, Forgetting 60min=1hr, Rushing conversions, Counting days one by one.

Geometry
Area & Perimeter

Geometry questions reward those who can read diagrams, not just calculate. If you can see the structure of a shape, its edges, its interior, and its hidden relationships, you will solve faster and with fewer mistakes.

This chapter trains your visual brain for selective and OC geometry. Some questions ask you to calculate. The best ones ask you to see.

Your Memory Anchor

Perimeter walks around. Area fills inside.
Add for edges. Multiply for space.
"Overlaps don't count twice.

This anchor solves most beginner geometry mistakes:

- ✓ Perimeter → outside boundary
- ✓ Area → space covered
- ✓ Add lengths for edges
- ✓ Multiply dimensions for space
- ✓ Shared or overlapping parts count once only

Keep this in mind whenever a diagram looks confusing.

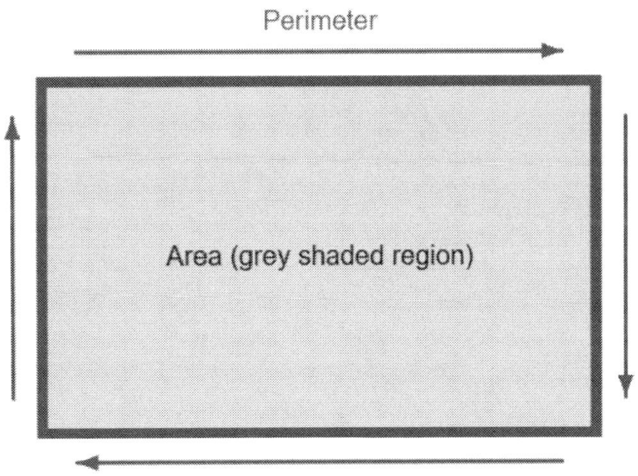

The Two Fundamental Concepts

Perimeter:
Perimeter is the total distance around the outside of a shape. To find perimeter, add only the outer edges.

Area:
Area is the amount of space inside a shape. To find area, measure how much surface is covered.

From the anchor: "***Perimeter walks around. Area fills inside.***"

Square

A **square** is a special type of quadrilateral.

It has:

- ✓ 4 equal sides
- ✓ 4 right angles (each angle is **90°**)

Area

The amount of space inside the square:

Area = s^2

Perimeter

The total distance around the square:

Perimeter = $4s$

Worked Example

Side = 6 cm

Area = $6 \times 6 = 36\ \text{cm}^2$

Perimeter = $4 \times 6 = 24$ cm

Visual Tip

A square is just a special rectangle with equal sides.

Rectangle

A rectangle has **4 sides**, but only the **opposite sides are equal**. Rectangles are very common and are used in many bigger shapes.

Area = length × width

Perimeter = 2(length + width)

Worked Example

Rectangle 8 cm by 3 cm

Area = $8 \times 3 = 24\ \text{cm}^2$

Perimeter = $2(8 + 3) = 22$ cm

Triangle

A triangle has 3 sides and 3 angles.
The angles in any triangle always add up to: **180°**

Perimeter = a + b + c

Area = ½ × base × height

Important

The height must be **perpendicular** to the base, meaning it goes **straight up at a right angle**. It is not always one of the sides of the triangle.

💡 **Visual Meaning**

*A **triangle** is always half of a **rectangle** with the same base and height.*

*A **square** is a special type of rectangle, so the rule works there too.*

Worked Example

Base = 10 cm, height = 4 cm

Area = $\frac{1}{2} \times 10 \times 4 = 20\ \text{cm}^2$

Circle

It is a perfectly round shape where every point on the edge is the same distance from the centre. The word circle comes from the Latin word: circulus.

Meaning:

→ A small ring

→ Something round

→ Something that goes around

So a circle is called a circle because it forms a continuous round loop, like a ring.

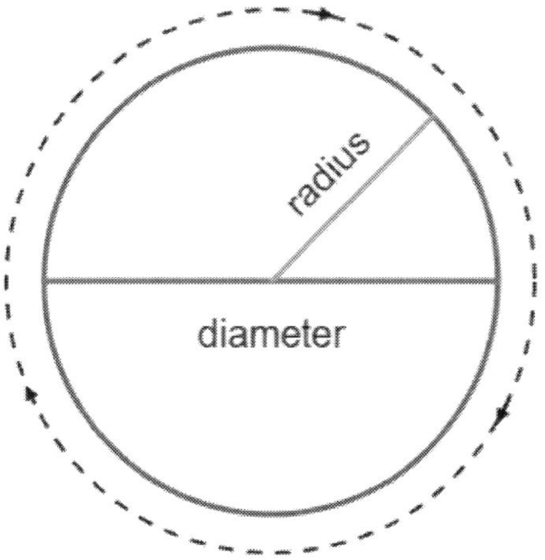

Key Parts of a Circle:

1. Radius ***r***: distance from the centre to the edge.
2. Diameter ***d***: distance across the circle through the centre.

$$\boldsymbol{d} = 2r$$

Perimeter (Circumference)

The distance around the circle is called the circumference:

$$C = 2\pi r$$

$$C = \pi d$$

Area

The amount of space inside the circle:

$$A = \pi r^2$$

The Constant π (Pi) $\boldsymbol{\pi \approx 3.14}$

Selective school students may also use $\boldsymbol{\pi} \approx \frac{22}{7}$ *for more accuracy.*

Worked Example

Radius = 5 cm

Area = $\pi \times 5^2 = 25\pi = 25 \times 3.14 = 78.5\ \text{cm}^2$

Circumference = $2\pi \times 5 = 10\pi = 10 \times 3.14 = 31.4$ cm

The FAST CHECK Method

Before doing any working, ask yourself these three questions:

- ✓ Are they asking for perimeter or area?
- ✓ Are there hidden sides or missing lengths?
- ✓ Are any sides shared or overlapping?

Composite Shapes

A composite shape is a shape made by joining two or more simple shapes, such as:

- ✓ Squares
- ✓ Rectangles
- ✓ Triangles
- ✓ Parts of circles

Think of it like a shape made from **building blocks stuck together**.

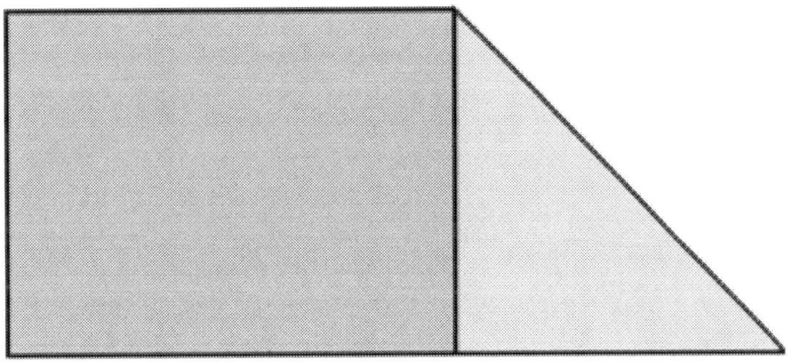

Finding the Area of Composite Shapes

Step-by-step method:

1. Break the shape into smaller shapes you recognise
2. Find the area of each part
3. Add them together

⚠ **Important**

Don't try to guess one big formula. Always split it up!

Worked Example

Question: A composite shape is made from two rectangles. Rectangle 1 has dimensions 6 cm × 4 cm, and Rectangle 2 has dimensions 3 cm × 4 cm. Find the total area of the composite shape.

Steps:

Step 1: Area of Rectangle 1

$6 \times 4 = 24 \text{ cm}^2$

Step 2: Area of Rectangle 2

$3 \times 4 = 12 \text{ cm}^2$

Step 3: Add them

$24 + 12 = 36 \text{ cm}^2$

✓ **Answer:** 36 cm^2

Finding the Perimeter of Composite Shapes

To find the perimeter:

Walk around the outside edge only. Imagine an ant walking around the shape's boundary.

📌 **Do NOT count:**

- Inside lines.
- Shared edges between shapes.

Only count the outer boundary.

Perimeter Shortcut (Exam Trick!)

If two shapes are joined together:

$$\text{Perimeter of joined shape} = (\text{sum of perimeters}) - 2 \times (\text{shared edge})$$

Why subtract?

Because the shared edge is inside, not on the outside.

Quick Memory Tips

- ✓ *Area = Split + Solve + Add*
- ✓ *Perimeter = Outside Only*
- ✓ *Shared edges don't count*

Missing Side Lengths in a Rectangle

A rectangle has 4 sides.
The perimeter is the total distance around the rectangle.

Since a rectangle has:

- → 2 equal lengths
- → 2 equal widths

The perimeter always includes each side **twice**.

So, if one side is missing, we can use the perimeter to find it.

Rule (Important to Remember)

- ✓ **Opposite sides of a rectangle are equal**

  Top = Bottom

  Left = Right

- ✓ **Perimeter Formula**

  $$\mathbf{P = 2L + 2W}$$

  Or

  $$\mathbf{P = 2(L + W)}$$

Steps to Find the Missing Side

Step 1: Divide the perimeter by 2

We **divide by 2** because the perimeter counts two lengths and two widths.
So dividing removes the "double counting" and gives:

$P \div 2 = L + W$
This means we now have the total of one length + one width.

Step 2: Subtract the known side

(L + W) - "known side" = "missing side"

Quick Memory Tips

Perimeter ÷ 2 → Subtract known side → Missing side

Worked Example

Perimeter = 28 cm, **one side =** 5 cm

Step1: 28 ÷ 2 = 14

Step2: 14 – 5 = 9 cm

Rearranging Shapes

→ **Rearranging** does not change area as long as there are no gaps and no overlaps.

→ **Same pieces** → same space → same area.

→ But **perimeter** may change, because the outside boundary changes.

Overlapping Shapes

When shapes touch or overlap:

Common Traps

- ☒ Do NOT double-count shared sides for perimeter.
- ☒ Do NOT count overlapping areas twice.

Rules:

- ✓ Shared edges are not perimeter.
- ✓ Overlapping regions count once for area.

Multiple-Choice Shortcut: "Boundary Walk"

When asked: Which perimeter is correct?

1. Trace the outside with your finger.
2. Ignore internal lines.
3. Add only the edges you touched.

This turns a confusing picture into a 5-second answer.

Try It Yourself

1. A square has side 7 cm. What are its area and perimeter?
2. A rectangle is 5 cm by 4 cm. What is its area?
3. A triangle has base 10 cm and height 6 cm. What is its area?
4. A circle has radius 3 cm. Find its area (use $\pi \approx 3.14$).
5. Two shapes are rearranged using the same pieces with no overlaps. Which has the greater area?
6. A composite shape has a shared internal side. Should that side be included in the perimeter? Why?

Key Takeaway

Perimeter walks around. Area fills inside.

Add for edges. Multiply for space.

Overlaps don't count twice.

You don't need long formulas. Read the picture. Think visually.

Avoid traps. Solve with confidence

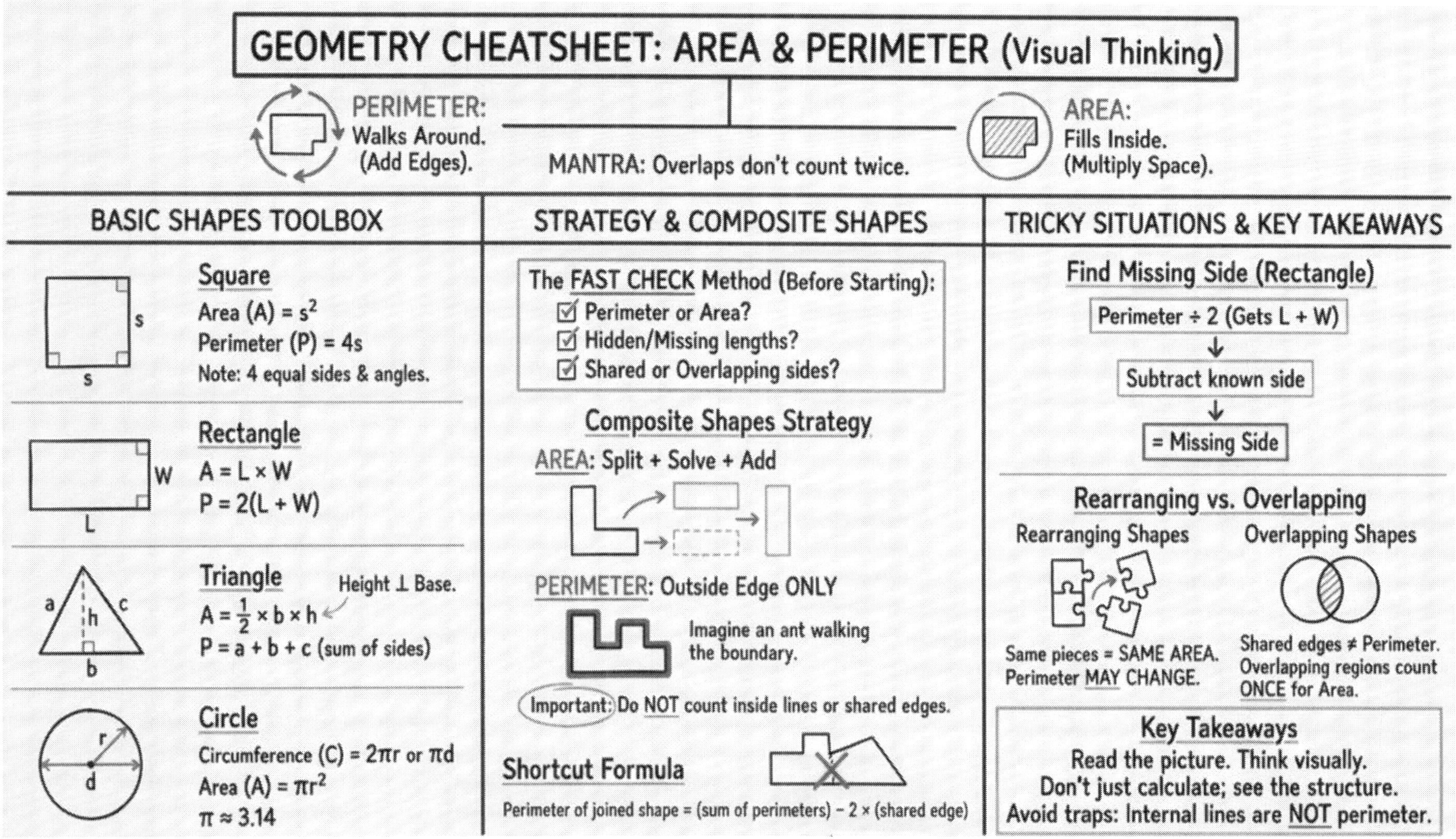
GEOMETRY CHEATSHEET: AREA & PERIMETER (Visual Thinking)
PERIMETER: Walks Around. (Add Edges).
AREA: Fills Inside. (Multiply Space).
MANTRA: Overlaps don't count twice.
BASIC SHAPES TOOLBOX
Square
Area (A) = s²
Perimeter (P) = 4s
Note: 4 equal sides & angles.
Rectangle
A = L × W
P = 2(L + W)
Triangle
Height ⊥ Base.
A = 1/2 × b × h
P = a + b + c (sum of sides)
Circle
Circumference (C) = 2πr or πd
Area (A) = πr²
π ≈ 3.14
STRATEGY & COMPOSITE SHAPES
The FAST CHECK Method (Before Starting):
Perimeter or Area?
Hidden/Missing lengths?
Shared or Overlapping sides?
Composite Shapes Strategy
AREA: Split + Solve + Add
PERIMETER: Outside Edge ONLY
Imagine an ant walking the boundary.
Important: Do NOT count inside lines or shared edges.
Shortcut Formula
Perimeter of joined shape = (sum of perimeters) − 2 × (shared edge)
TRICKY SITUATIONS & KEY TAKEAWAYS
Find Missing Side (Rectangle)
Perimeter ÷ 2 (Gets L + W)
Subtract known side
= Missing Side
Rearranging vs. Overlapping
Rearranging Shapes
Overlapping Shapes
Same pieces = SAME AREA. Perimeter MAY CHANGE.
Shared edges ≠ Perimeter. Overlapping regions count ONCE for Area.
Key Takeaways
Read the picture. Think visually.
Don't just calculate; see the structure.
Avoid traps: Internal lines are NOT perimeter.

Geometry

Angles, Symmetry & Transformations

Geometry follows rules. Some angles always add to 180°. Some shapes love mirrors. Some figures spin and still look the same. If you can recognise these shape rules, you will solve faster, eliminate wrong answers, and avoid unnecessary working. This chapter gives you high-value shortcuts for geometry questions.

Your Memory Anchor

On a line one eight zero

Round a point three six zero

Opposites match equal all

Z and F are same each call

C goes straight add to one eight zero

Triangle one eight zero

Mirror flips rotation spins

Spot the rule — that's how you win.

What is an Angle?

An **angle** is one of the most important ideas in geometry.

It is formed when two rays (or straight lines) meet at a common endpoint. That endpoint is called the **vertex**.

An angle shows us how much one line turns away from another, and we measure angles in **degrees (°)**.

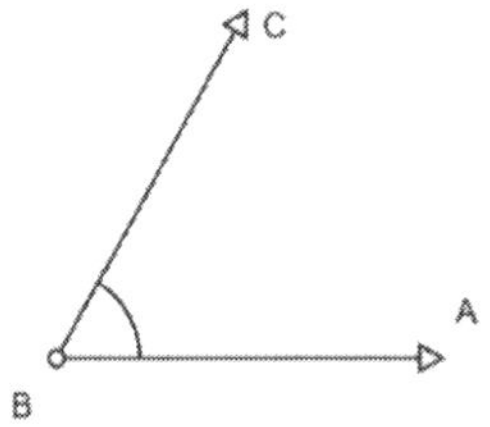

- **Vertex: B** (the point where the two rays **BA** and **BC** meet).
- **Angle:** The opening between BA and BC. **Angle Name:** $\angle ABC$

Example

If two lines open up only a little, the angle is **small**.

If the lines open wider, the angle becomes **larger**.

So, angles help us describe the amount of turning between lines.

Angle Types You Should Recognise

Include:

- ✓ Acute (< 90°)
- ✓ Right (= 90°)
- ✓ Obtuse (90–180°)
- ✓ Reflex (> 180°)

Key Angle Rules You Must Know

Angle rules appear frequently in selective exams and knowing them can help you solve problems much faster.

Angles on a Straight Line

When angles lie next to each other on a **straight line**, they are called **supplementary**.

This means their total is always: **180°**

So, for two angles on a straight line: $\angle \mathbf{a} + \angle \mathbf{b} = \mathbf{180}^\circ$

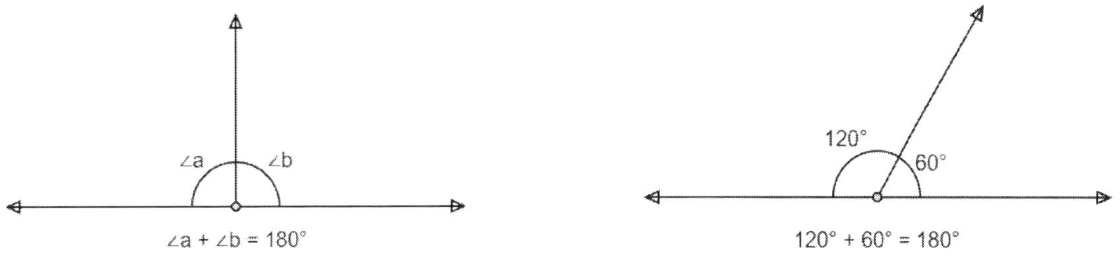

Figure: Angles on a Straight Line (Supplementary Angles)

Angles Around a Point

When several angles meet around a **single point**, they form a **full turn** (a complete rotation).

Their total is always:

$$360^\circ$$

So, for angles around a point:

$$\angle a + \angle b + \angle c + \cdots = 360^\circ$$

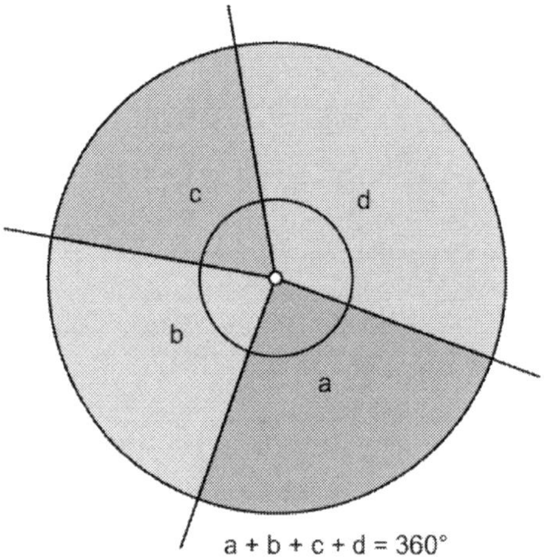

Figure: Angles Around a Point

Vertically Opposite Angles

When two straight lines intersect (cross), they form four angles.

The angles that are directly opposite each other are called vertically opposite angles, and they are always equal. **Opposite angles are equal**.

Worked Example

Question: Two straight lines cross. One angle measures 42°. What is the angle directly opposite it?

Think smart:

Vertically opposite angles are always equal.

So, Opposite angle = **42°**

✓ **Answer: 42°**

Parallel Line Angles

When a **transversal** (a straight line) cuts across two **parallel lines**, several angles are formed.

Even though the diagram may look complicated, the angles follow very predictable patterns. These patterns are easy to recognise because they look like familiar letters:

- → Z-shape
- → F-shape
- → C-shape

Learning these shapes helps you solve angle questions quickly.

Alternate Angles — Z Shape

What it looks like

The two angles lie on opposite sides of the transversal, forming a **Z pattern**.

Alternate angles are always equal.

$$\angle a = \angle a$$

If you can draw a **Z** through the angles, they match.

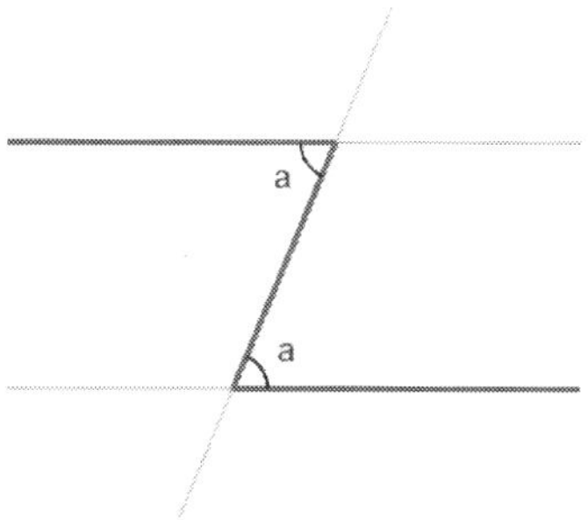

Figure: Z pattern

Corresponding Angles — F Shape

What it looks like

The angles are in the same position at each intersection, forming an **F pattern**.

Corresponding angles are always equal.

$$\angle a = \angle a$$

If the angles sit in the same corner, they are the same.

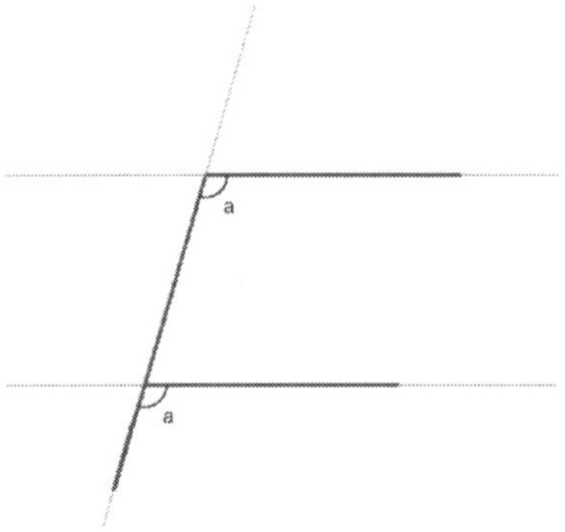

Figure: F pattern

Co-Interior Angles — C Shape

What it looks like

Both angles lie inside the parallel lines, on the same side of the transversal, forming a C pattern.

Co-interior angles add up to 180°.

$$\angle \boldsymbol{a} + \angle \boldsymbol{b} = \mathbf{180}^{\circ}$$

If you can trace a C around them, they make a straight line total.

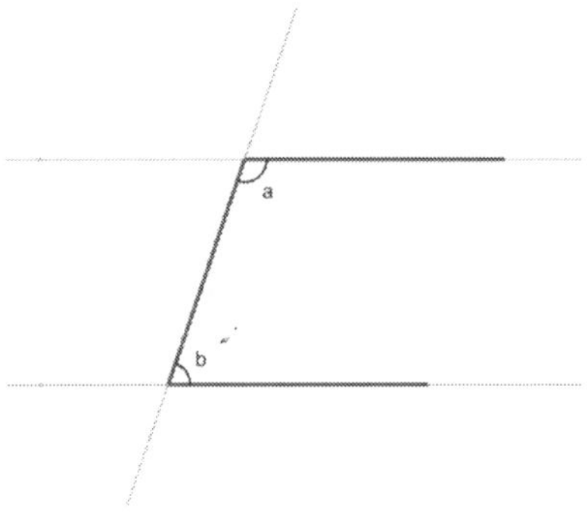

Figure: C pattern

Quick Shape Reminder

Shape	Angle Type	Rule
Z	Alternate angles	Equal
F	Corresponding angles	Equal
C	Co-interior angles	Add to 180°

Worked Example

Question: Two parallel lines are cut by a transversal. One angle is 124°. Find x in the Z position.

Think smart:

Z-angles are equal

So, **x = 124**

✓ **Answer: 124**

Question: A C-shaped pair of angles has one angle of 36°. Find the other.

Think smart:

C-angles add to 180°

x = 180 − 36

= **144°**

✓ **Answer: 144°**

Polygon Angle Rule

One of the most useful geometry facts is that all polygons follow the same angle rule.

What is a Polygon?

A polygon is a closed shape made only from straight sides.

That means you can draw it using only a ruler, with no curves.

Examples of polygons:

- ✓ Triangle (3 sides)
- ✓ Square (4 sides)
- ✓ Pentagon (5 sides)
- ✓ Hexagon (6 sides)

📌 *A circle is not a polygon because it has no straight sides.*

The Polygon Angle Formula

To find the total of all the inside angles of any polygon, use:

$$(\boldsymbol{n} - \mathbf{2}) \times \mathbf{180}^{\circ}$$

Where: **n** = the number of sides

Triangle (3 sides)

$$(\mathbf{3} - \mathbf{2}) \times \mathbf{180}^{\circ} = \mathbf{180}^{\circ}$$

So the angles inside a triangle always add to: $\mathbf{180}^{\circ}$

Quadrilateral (4 sides)

$$(\mathbf{4} - \mathbf{2}) \times \mathbf{180}^{\circ} = \mathbf{360}^{\circ}$$

So the angles inside a quadrilateral add to: $\mathbf{360}^{\circ}$

Pentagon (5 sides)

$$(\mathbf{5} - \mathbf{2}) \times \mathbf{180}^{\circ} = \mathbf{540}^{\circ}$$

So the angles inside a pentagon add to: $\mathbf{540}^{\circ}$

Why Does This Work?

Here's the secret:

Every polygon can be split into triangles.

- Triangle → 1 triangle
- Quadrilateral → 2 triangles
- Pentagon → 3 triangles

The number of triangles is always: $\boldsymbol{n} - \mathbf{2}$

Each triangle has angles adding to: $\mathbf{180}^{\circ}$

So, we multiply: $(\boldsymbol{n} - \mathbf{2}) \times \mathbf{180}^{\circ}$

Exterior Angle Rule

For any polygon:

Sum of exterior angles = 360°

This helps:

- ✓ Regular polygon angle problems
- ✓ Missing angle diagrams

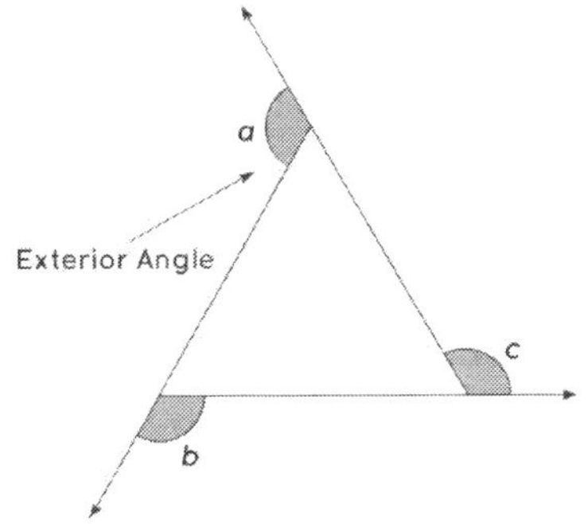

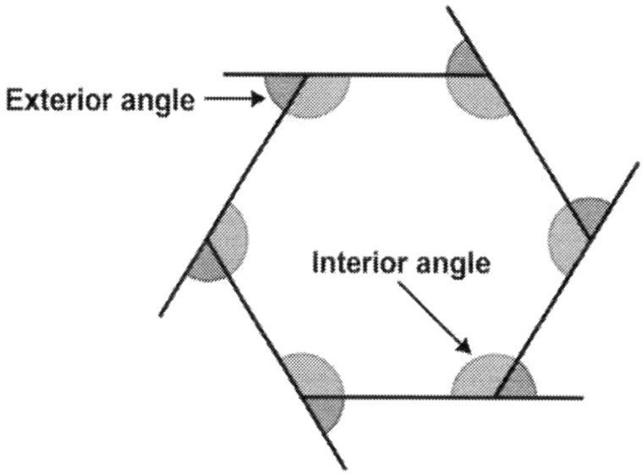

Worked Example

Question: A hexagon has **6 sides**. Find the total sum of its **interior angles**.

Steps:

Step 1: Use the Polygon Formula

$$(\boldsymbol{n} - \mathbf{2}) \times \mathbf{180^\circ}$$

Step 2: Substitute $\boldsymbol{n} = \mathbf{6}$

$$(\mathbf{6} - \mathbf{2}) \times \mathbf{180^\circ}$$

Step 3: Simplify

$$\mathbf{4} \times \mathbf{180^\circ} = \mathbf{720^\circ}$$

✓ **Answer:** $\mathbf{720^\circ}$

Memory Anchor Scan

Before doing any working, scan the diagram:

- ✓ On a straight line? → **180°** → On a line one eight zero.
- ✓ Around a point? → **360°** → Round a point three six zero.
- ✓ Opposite angles? → **Equal** → Opposites match equal all.
- ✓ Parallel lines with Z or F? → **Same** → Z and F are same each call.
- ✓ Parallel lines with C? → **180°** → C goes straight add to one eight zero.
- ✓ Triangle? → **Total 180°** → Triangle one eight zero.

Geometry Fast Scan

Checklist style:

- ✓ Straight line → 180°
- ✓ Around a point → 360°
- ✓ Opposite angles equal
- ✓ Z/F angles equal
- ✓ C angles add to 180°
- ✓ Triangle = 180°
- ✓ Symmetry = equal parts

Symmetry

Symmetry is one of the easiest geometry topics once you know what to look for.

A shape is symmetrical when it looks the same on both sides.

Simple Idea

A shape has symmetry if you can fold it in half and both parts match perfectly.

- → One half is a mirror image of the other.
- → The fold line is called the line of symmetry.

Lines of Symmetry

A line of symmetry is a line that divides a shape into two equal halves. Each half is a perfect mirror image of the other.

Fold Test:
If you fold the shape along the line and both sides line up exactly, that line is a line of symmetry.

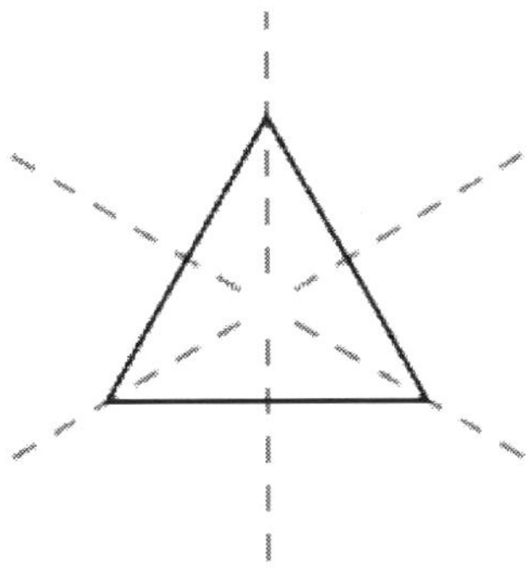

Equilateral triangle
3 lines of Symmetry

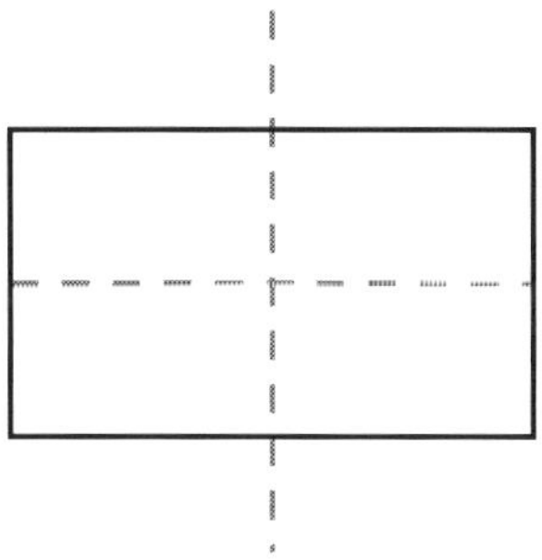

Rectangle
2 lines of Symmetry

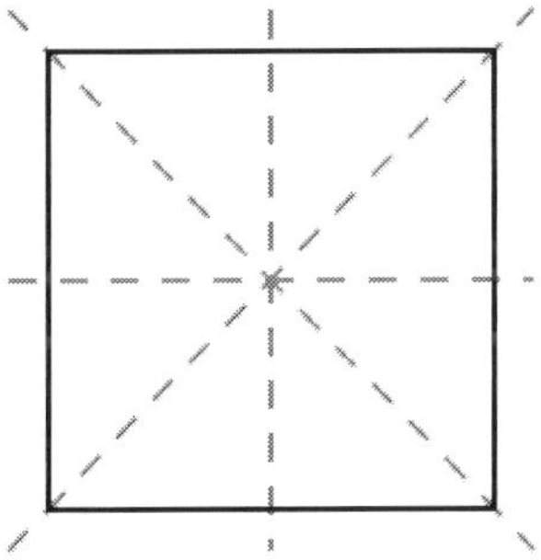

Square
4 lines of Symmetry

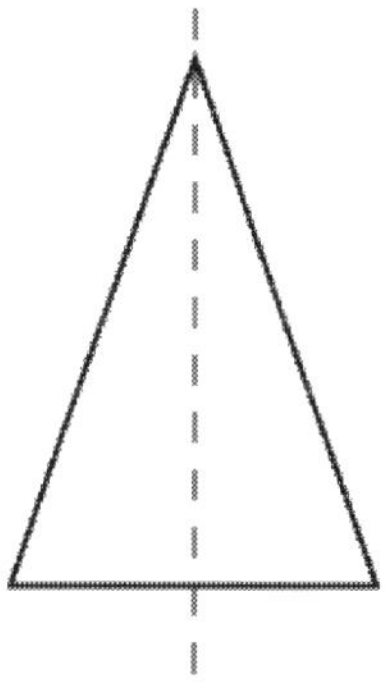

Isosceles triangle
1 line of Symmetry

Figure: Symmetry Examples

Common Shapes and Their Lines of Symmetry

Shape	Lines of Symmetry
Square	4
Rectangle	2
Equilateral triangle	3
Isosceles triangle	1
Scalene triangle	0
Regular hexagon	6
Circle	Infinite

Rotational Symmetry

A shape has rotational symmetry if it can be turned around its centre and still look exactly the same before completing a full 360° turn.

The order of rotational symmetry tells us how many times the shape matches itself in one full rotation.

Examples of Rotational Symmetry

Shape	Order	Rotation Interval
Square	4	Every 90°
Rectangle	2	Every 180°
Equilateral triangle	3	Every 120°
Regular pentagon	5	Every 72°
Irregular shape	1	Only at 360°

💡 Symmetry Fast Check

✗ ***Not symmetrical if:*** *sides are different lengths, angles don't match, it only looks the same after a full 360° turn.*

✓ ***Symmetrical if:*** *you can fold it exactly in half, or rotate it and it matches before completing a full turn.*

Transformations

In selective school geometry, transformations are not tested as simple "flip a shape" questions.

Instead, they appear as hidden shortcuts inside harder diagram problems.

Big Idea

Transformations help you quickly spot:

- ✓ Equal angles.
- ✓ Equal sides.
- ✓ Symmetry patterns.
- ✓ Repeated shapes.

Selective students don't do more work, they spot the rule first.

Types of Transformations

- ✓ Translation → slide
- ✓ Reflection → flip
- ✓ Rotation → turn
- ✓ Enlargement → resize (scale factor)

Reflection (Mirror)

A reflection flips a shape across a line called the mirror line. It is like holding the shape up to a mirror.

Key Rule (Congruent Shapes)

When a shape is reflected, the new shape is congruent to the original.

Congruent means:

- ✓ Exactly the same size
- ✓ Exactly the same shape
- ✓ Matching sides stay equal
- ✓ Matching angles stay equal

The only difference is that the shape faces the opposite way.

Reflection ***changes direction, not measurements.***

Worked Example

Question: A triangle is **reflected** across a mirror line. One angle in the original triangle is: **38°**. What is the **matching angle** in the **reflected triangle**?

Think smart:

Step 1: Reflections create congruent shapes. So angles do not change.

Step 2: Matching angles stay equal **38°**

✓ **Answer: 38°**

Rotation (Symmetry Angle)

A rotation turns a shape around a fixed point called the centre of rotation.

It is like spinning a shape around a pin.

Key Rule

Rotations do not change size or shape. So the rotated shape is also congruent to the original.

Rotation + Symmetry Link

Selective exams often combine rotation with symmetry.

If a shape has rotational symmetry of order n:

$$\textbf{Smallest rotation angle} = \frac{360^\circ}{n}$$

Worked Example

Question: A shape matches itself **5 times** in one **full turn**. What is the **smallest angle** it can rotate and still look the same?

Think smart:

Step 1: Order $\boldsymbol{n} = \mathbf{5}$

Step 2: Divide the full turn $360^\circ \div 5 = 72^\circ$

✓ **Answer: 72°**

Enlargement (Scale Factor Idea)

An enlargement changes the **size** of a shape but keeps the **shape exactly the same**.

Think of it like zooming in or out on a photo.

- ✓ Same shape.
- ✓ Same angles.
- ✓ Side lengths change proportionally.

The number that tells us how much the shape grows or shrinks is called the **scale factor**.

Key Idea

- ✓ Scale factor greater than 1 → shape gets bigger
- ✓ Scale factor between 0 and 1 → shape gets smaller
- ✓ Scale factor of 1 → shape stays the same size

Even though the size changes, the shape remains **similar** to the original.

That means:

- ✓ Angles stay equal
- ✓ Side lengths change in the same ratio

Reasoning From Diagrams

Before doing any calculations, pause and ask:

- ✓ Are any angles **equal** because of **symmetry**?
- ✓ Can you spot a **Z, F, or C** angle pattern?
- ✓ Has the shape been **reflected** or rotated?
- ✓ Do lengths or angles **stay the same** because shapes are **congruent**?

Try It Yourself

1. Two angles on a straight line are 72° and x. Find x.
2. Two lines cross. One angle is 55°. What is the opposite angle?
3. Two parallel lines form a Z shape. One angle is 118°. Find the matching angle.
4. A C-angle pair has one angle of 39°. What is the other?
5. A triangle has angles of 48° and 67°. Find the third angle.
6. How many lines of symmetry does a square have?
7. What is the order of rotational symmetry of a rectangle?
8. A shape is reflected. What changes: size, shape, or direction?

Key Takeaway

On a line one eight zero

Round a point three six zero

Opposites match equal all

Z and F are same each call

C goes straight add to one eight zero

Triangle one eight zero

Mirror flips rotation spins

Spot the rule — that's how you win.

GEOMETRY — ANGLES • SYMMETRY • TRANSFORMATIONS

Spot the rule. Eliminate fast. Solve with confidence.

1 Angle Basics

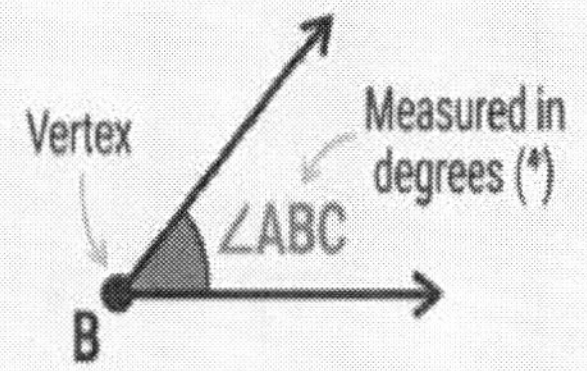

2 Core Angle Totals

On a line → 180°

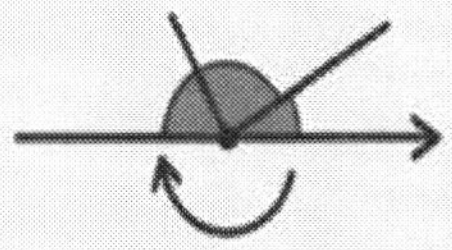

Around a point → 360°

Opposites → Equal

3 Parallel Line Patterns

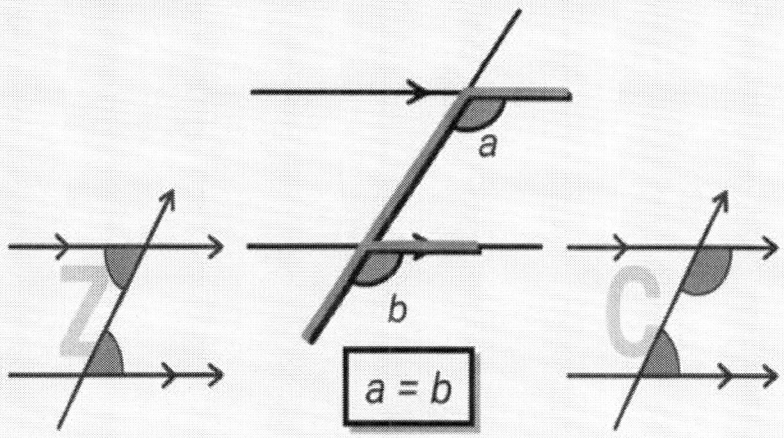

Z Pattern
Z → Alternate → Equal

F Pattern
F → Corresponding → Equal

C Pattern
C → Co-interior → Add to 180°

4 Polygon Angle Rule

$$(n - 2) \times 180°$$

n = number of sides

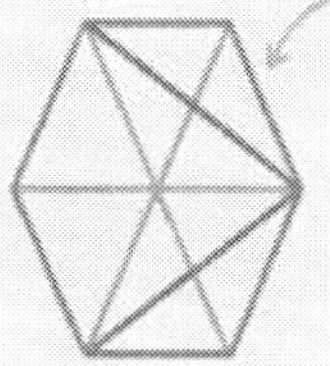

Polygons form (n – 2) triangles

Triangle → 180°
Quadrilateral → 360°
Pentagon → 540°

5 Lines of Symmetry

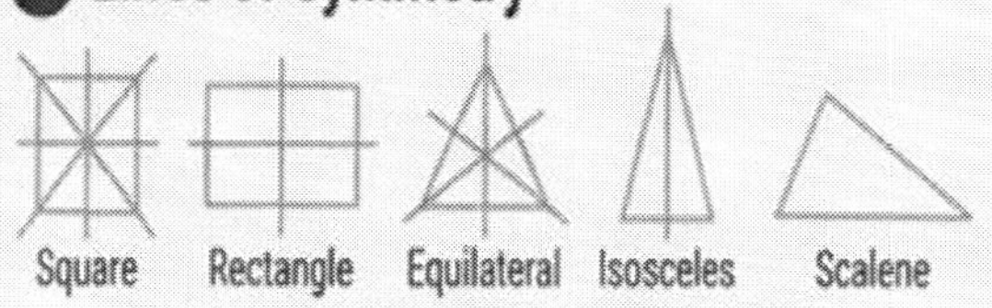

Lines of Symmetry → Fold Test
"Both halves match exactly."

6 Rotational Symmetry

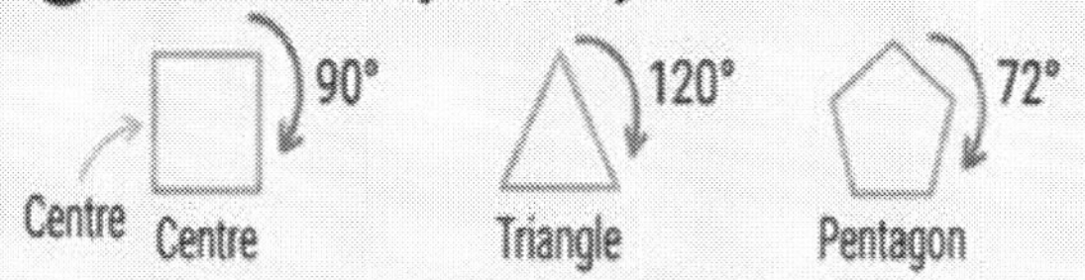

Smallest angle = 360° ÷ n
Order = number of matches in one full turn

7 Transformations

Reflection → Flip
- Same size
- Same angles
- Direction changes

Rotation → Spin
- Same size
- Same angles
- Orientation changes

Bottom Memory Anchor Strip

Memory Rhyme					
	On a line one eight zero Round a point three six zero	Opposites match equal all Z and F are same each call	C goes straight add ad to one eight zero	Triangle one eight zero Mirror flips rotation spins	Spot the rule – that's how you win

3D Shapes & Nets

See It. Count It. Fold It. Don't Guess.

Three-dimensional shapes are everywhere, including boxes, dice, buildings, cans, pyramids, and tents.

In OC and Selective School exams, they appear as:

- ✓ Counting problems involving faces, edges, and vertices
- ✓ Visual reasoning questions
- ✓ Folding and net puzzles
- ✓ Shadow and view questions

If you can see the structure clearly, you do not need trial and error.

This chapter gives you the shortcuts and the understanding behind them.

Your Memory Anchor

Faces are FLAT.

Edges are LINES.

Vertices are CORNERS.

The Three Features You Must Know

Faces

Faces are the flat surfaces of a 3D shape.

- ✓ A cube has 6 faces.
- ✓ A triangular prism has 5 faces.

From the anchor: "**Faces are flat**."

Edges

Edges are where two faces meet, forming straight or curved lines.

- ✓ A cube has 12 edges.
- ✓ A cylinder has 2 curved edges.

From the anchor: "**Edges are lines.**"

Vertices

Vertices are corner points where edges meet.

- ✓ A cube has 8 vertices.
- ✓ A cone has 1 vertex.

From the anchor: "**Vertices are corners.**"

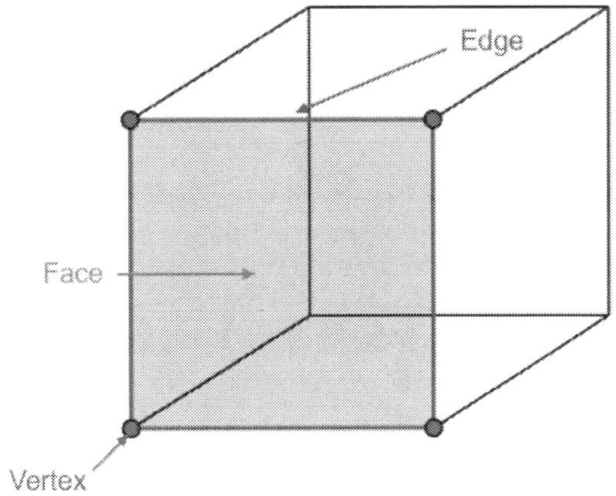

Fast Check (5 Seconds)

Before answering, ask yourself:

- ✓ Am I counting faces (flat surfaces)?
- ✓ Am I counting edges (lines)?
- ✓ Am I counting vertices (corners)?

Most mistakes come from counting the wrong feature.

Rotation Doesn't Change Structure

A shape can be tilted, rotated, or viewed differently, but its number of faces, edges, and vertices never changes.

Tip

- ✓ If it looks unfamiliar, imagine rotating it back.
- ✓ This prevents panic when shapes appear angled.

Know Your Solids

Shape	Faces	Edges	Vertices	Key Feature
Cube	6	12	8	All faces squares
Rectangular Prism	6	12	8	Opposite faces equal rectangles
Triangular Prism	5	9	6	Two triangular bases
Square Pyramid	5	8	5	One square base + four triangles
Cylinder	2 flat + 1 curved	2 curved	0	Circular ends
Cone	1 flat + 1 curved	1 curved	1	Pointed top
Sphere	1 curved	0	0	Fully round

Structure Check Rule

Many solid shapes follow a useful pattern linking faces, edges, and vertices. Think of this as a **quick geometry self-check** during exams.

Euler's Rule

Faces + Vertices – Edges = 2

Use it to:

- ✓ Check your counting.
- ✓ Find a missing value.
- ✓ Catch careless mistakes.

If the result isn't **2**, recheck your numbers.

Worked Example

Question: A solid has 6 faces and 8 vertices. How many edges?

$6 + 8 - \text{Edges} = 2$
$14 - \text{Edges} = 2$
$\text{Edges} = \mathbf{12}$

✓ **Answer:** 12. Structure confirms the answer.

Nets: Flat Shapes That Fold Into Solids

A net is a 2D pattern that folds into a 3D shape.

Cut → Fold → Build

If it can fold without overlap or gaps, it's a valid net.

👉 **The Golden Rule**

*A cube has **6 faces**, so every valid cube net must have exactly **6 squares**.*

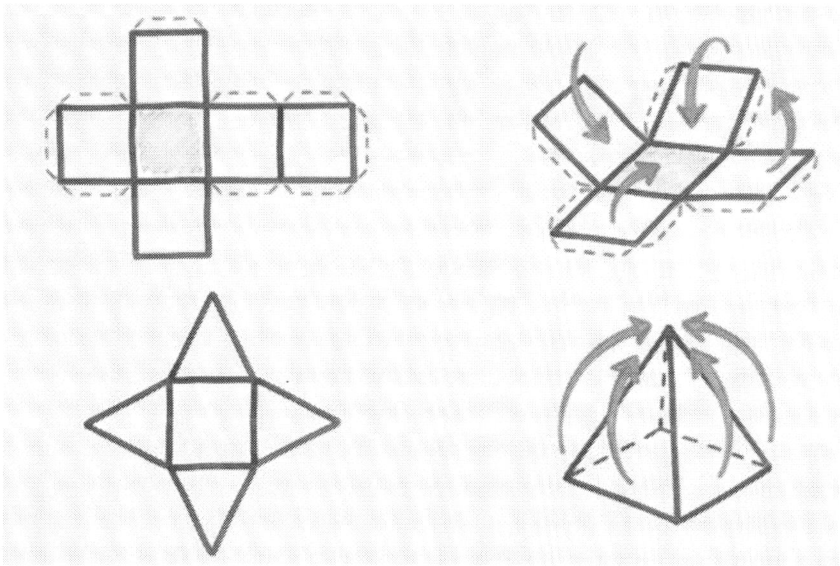

Valid Net Criteria

A net is valid if and only if:

- ✓ Faces are connected by full edges.
- ✓ It folds so that all faces meet neatly.
- ✓ There are no overlaps when folded.

A net is invalid if:

- ☒ Faces collide or overlap.
- ☒ There are gaps when folded.
- ☒ There are extra or missing faces.

Multiple-Choice Shortcut: "Net Check"

When asked "Which net makes a cube or prism?"

1. Count faces → does it match the solid?
2. Check connections → touching by full edges only
3. Imagine folding → will any faces overlap or leave gaps?
4. Eliminate quickly → test only what remains.

This turns a 40-second guessing question into a 10-second decision.

Worked Example

Question: Which net can fold to make a cube?

Think smart:

Option A → one face not connected → ✕

Option B → folding causes overlap → ✕

Option C → all 6 faces connected, no overlap → ✓

Option D → missing one face → ✕

✓ **Answer:** C — No guessing. Just structure.

Worked Example

Question: How many edges does a triangular prism have?

Think smart:

Count edges on each face:

Top triangle: 3 edges

Bottom triangle: 3 edges

Vertical edges joining them: 3 edges

Total:

$3 + 3 + 3 = 9$ edges

✓ **Answer:** 9 edges

Question: A shape has 6 faces, 12 edges, and 8 vertices. All faces are squares. What is it?

Think smart:

6 faces, 12 edges, 8 vertices → could be **cube** or **rectangular prism.**

All faces are **squares** → only one shape fits

Note: A rectangular prism can have some square faces, but cannot have all 6 faces square unless it is a **cube**.

✓ Answer: Cube.

Common Traps to Avoid

Common Mistakes

- ☒ Counting curved surfaces as edges or vertices.
- ☒ Mixing up faces and edges.
- ☒ Forgetting that nets must fold without overlap.
- ☒ Guessing shadows without checking flat faces.

Try It Yourself

1. How many vertices does a triangular prism have?
2. How many faces does a cylinder have?
3. Which feature is a corner point called?
4. Can this net make a cube? Why or why not?
5. A shape has 5 faces, 8 edges, and 5 vertices. What is it?
6. Which 3D shape could make a circular shadow?

   A) Cube B) Cylinder C) Pyramid D) Prism

Key Takeaway

Faces are FLAT.
Edges are LINES.
Vertices are CORNERS.

Don't guess. Don't overthink.
See the structure. Count logically.
Fold with confidence.

3D Shapes & Nets

See It. Count It. Fold It. Don't Guess.

Faces = FLAT | Edges ≠ LINES | Vertices ≠ CORNERS.

Faces

Flat surfaces

- Cube ~ 6
- Triangular prism ~ 5

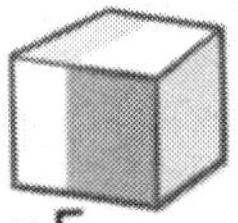

Edges

Where faces meet

- Cube ~ 12
- Cylinder ~ 2 (Curved)

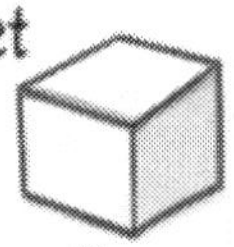

Vertices

Corner points

- Cube ~ 8
- Cone ~ 1

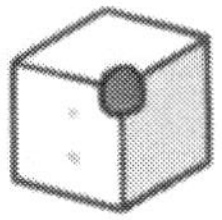

Nets: Flat → Fold → Solid

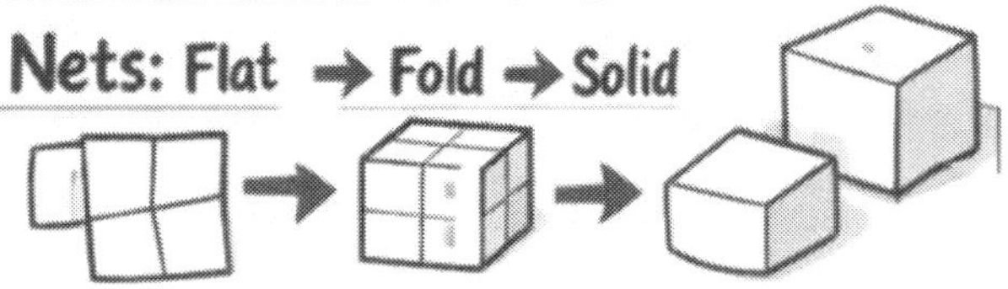

Know Your Solids

Shape	Faces	Edges	Vertices	Key Note
Cube	6	12	8	All faces
Rectangular Prism	6	12	8	Opposite faces equal
Triangular Prism	5	9	6	2 triangles + 3 rectangles
Square 7 Yramid	5	8	5	1 square - +1 triangles

Net Check (10 Seconds)

1. Count faces
2. Check connections
3. Visitualise fold
4. Eliminate quickly

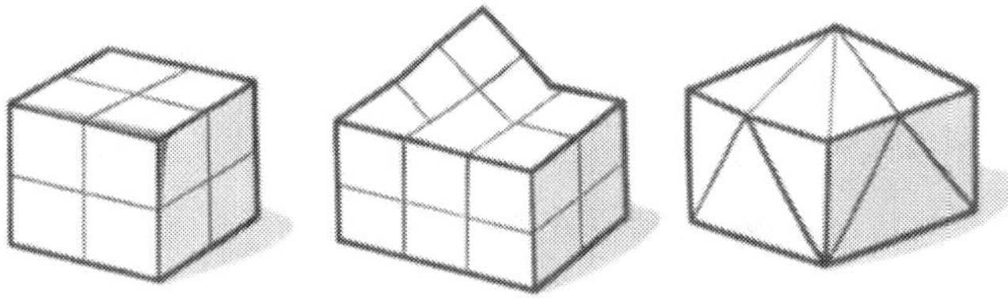

5-Second Check

✓ Counting faces (flat)?
✓ Counting edges (lines)?
✓ Counting vertices (cornes)?

Most mistakes = wrong feature counted.

Common Traps

- Counting curved surfaces
- Mixing up faces & edges
- Ignoring overlap in nets
- Guessing structure

See the structure. Count logically. Fold with confidence.

Charts & Graphs

Graphs tell stories. Some show winners and losers. Some show growth and decline. Some slice up a whole into neat portions.

If you can read the story quickly, you will:

- ✓ Solve faster.
- ✓ Spot traps instantly.
- ✓ Avoid careless mistakes.
- ✓ pick up easy marks with confidence.

In OC and Selective exams, graph questions are often free points if you read them correctly. This chapter turns looking at pictures into powerful problem solving.

Your Memory Anchor

Bars compare, lines show change.
Pie shows parts, tables arrange.
Read the labels, check the scale,
Spot the trend — you will not fail!

Graphs come in different types, and each one has a special job. Once you know what the graph is meant to do, the question becomes much easier.

Bar (Column) Graphs

A bar graph uses rectangular bars to compare amounts between groups. Bars stand side by side like competitors in a race.

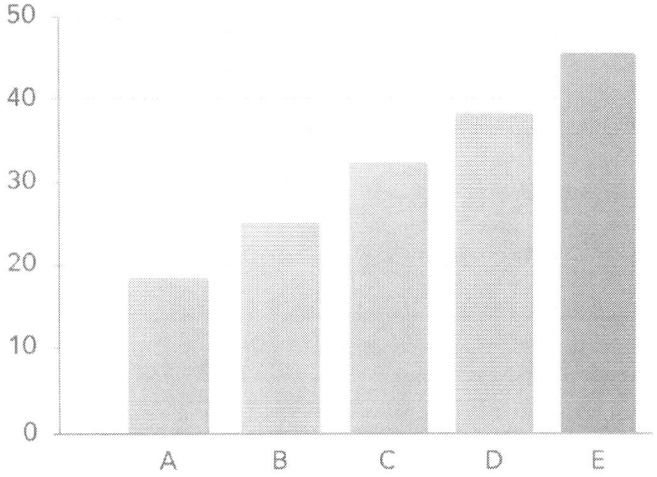

Best for: Comparing categories.

- ✓ *Who is tallest?*
- ✓ *Which month had the most sales?*
- ✓ *Which class scored the highest?*

Why Bar Graphs Are Easy Marks

- → Taller bar = bigger value
- → Shorter bar = smaller value
- → Great for "most" and "least" questions

Bar graphs are **visual comparisons**. Your brain sees taller versus shorter before it calculates anything. But this is important, your eyes can be tricked by scale. Never guess. Always confirm the number.

Quick Memory Tips

Bars are like a scoreboard. They compare winners and losers.

The Pencil Trace Technique

When asked for a value:

1. Place your pencil at the top of the bar.
2. Trace straight across to the axis.
3. Read the exact number from the scale.
4. Only then calculate if needed.

Your eyes guess. Your pencil doesn't.

Worked Example

Question: How many more students joined Science than the average number who joined Drama and Chess?

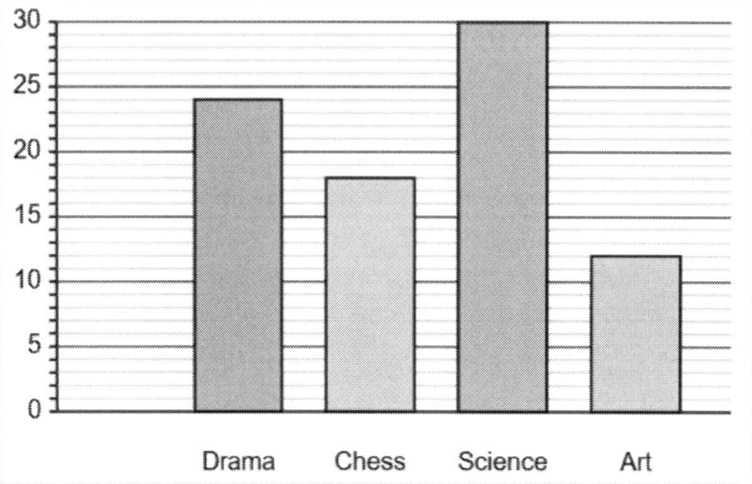

Steps:

Step 1: Average Drama and Chess:

24 + 18 = 42
42 ÷ 2 = 21

Step 2: Compare Science with this average:

30 − 21 = **9**

✓ **Answer: 9**

Line Graphs

A line graph joins points with a line to show change over time. Think of it like a journey: the line moves up, down, or stays flat.

Line graphs show change over time.

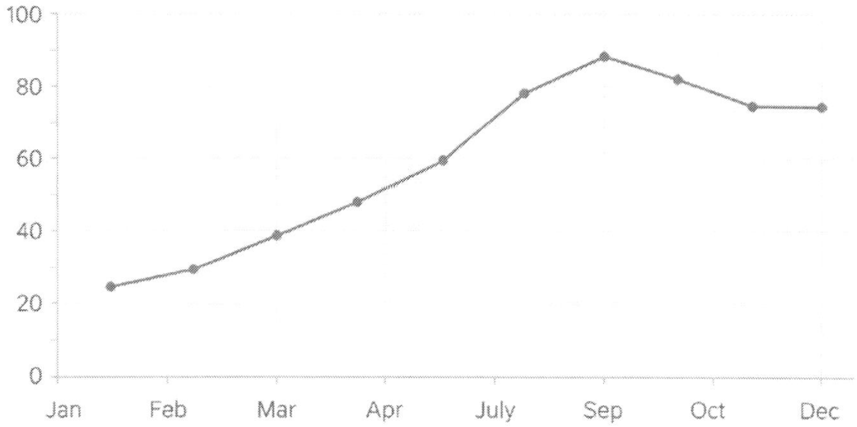

They answer:

- ✓ Is it increasing?
- ✓ Decreasing?
- ✓ Staying the same?
- ✓ By how much did it change?

Think of a line graph as a journey.

- ✓ Up = climbing
- ✓ Down = falling
- ✓ Flat = resting

Why Line Graphs Matter

Line graphs show movement.

Selective exams often test:

- ✓ Total increase.
- ✓ Greatest increase.
- ✓ Greatest drop.
- ✓ Overall change.

The Trend & Difference Method

When solving:

1. Identify start value.
2. Identify end value.
3. Subtract carefully.
4. Watch for drops (don't count them if question says "added").

Trend Reading Table

What You See	Meaning
↗ Upward line	Increasing
↘ Downward line	Decreasing
→ Flat line	No change

💡 Quick Memory Tips

Lines are like a heartbeat. They show change.

Worked Example

Question: How much water was added in total between 9 am and 6 pm?

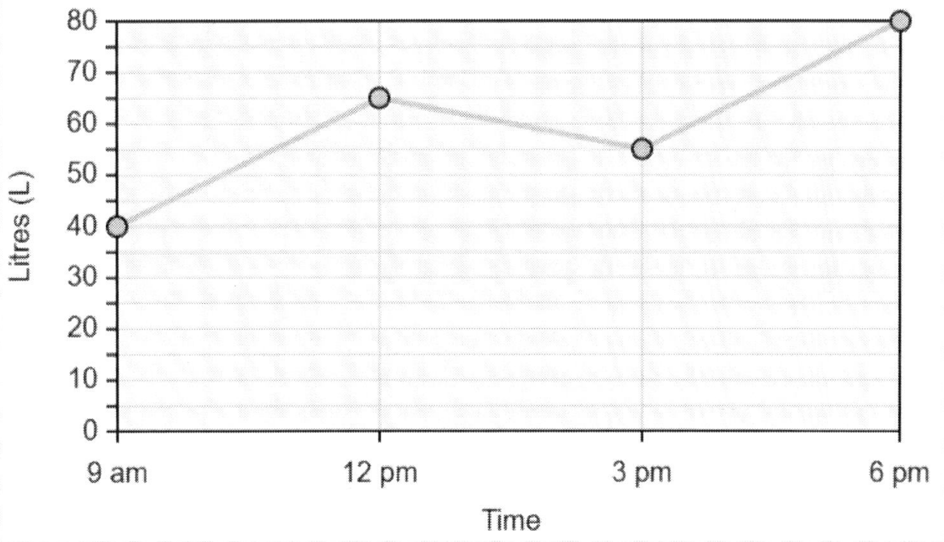

Steps:

Step 1: Increase from 9 to 12:

65 – 40 = 25 added

Step 2: Drop from 12 to 3:

65 – 55 = 10 removed (ignore for "added")

Step 3: Increase from 3 to 6:

80 – 55 = 25 added

Step 4: Total added:

25 + 25 = **50 L**

✓ **Answer: 50 L**

Pie Charts

A pie chart is a circle divided into slices. Each slice represents a fraction or percentage of the whole.

The full circle = **100%.**

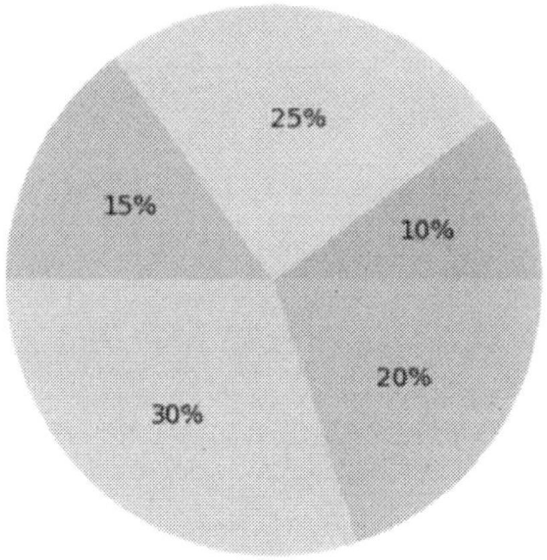

They answer:

- ✓ What percentage?
- ✓ What fraction?
- ✓ Which part is biggest?
- ✓ What is the missing slice?

Think of it like a pizza. Bigger slice = bigger portion.

Understanding Pie Charts Clearly

A pie chart always represents a **whole amount**. The full circle = **100% of the data.**

Each slice shows how much of the total belongs to one category.

So, every pie chart question is really asking one of three things:

- ✓ What fraction of the total is this part?
- ✓ What percentage of the total is this part?
- ✓ How many items does this percentage represent?

💡 The One Big Relationship

Every pie chart connects three things:

Part ↔ Whole ↔ Percentage

You move between them depending on what the question gives you.

If you know:

- ✓ *Part ÷ Whole = Fraction*
- ✓ *(Part ÷ Whole) × 100 = Percentage*
- ✓ *Number = Percentage × Total*

Finding the Fraction

Part ÷ Whole = Fraction

Use this when you know the actual numbers and want the fraction.

60 students out of 240 travels by bus. What fraction of the students travel by bus?

60 ÷ 240 = 1/4

So, one quarter travel by bus.

Finding the Percentage

(Part ÷ Whole) × 100 = Percentage

Use this when you know the numbers and want the percent.

60 students out of 240 travels by bus. What percentage of students travel by bus?

(60 ÷ 240) × 100 = 25%

So, 25% travel by bus.

Finding the Actual Number

Number = Percentage × Total

Use this when the chart gives a percentage and you need the actual number.

If 25% of 240 students travel by bus, how many students travel by bus?

25% × 240 = 60

So 60 students travel by bus

Worked Example

Question: The pie chart shows how 240 students travel to school. How many more students travel by car?

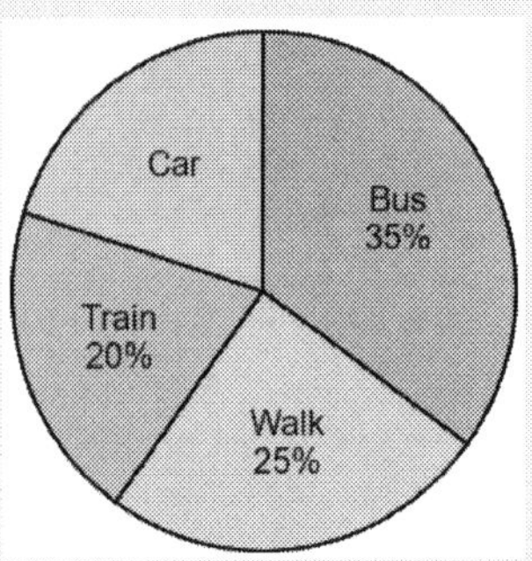

Steps:

Step 1: Find car percentage:

35 + 25 + 20 = 80%

100 – 80 = 20%

Step 2: Convert to numbers:

Car = 20% of 240 = **48**

✓ **Answer:** 48

Quick Memory Tips

Pie charts are pizzas. A bigger slice means a bigger portion.

Tables

Tables are neat, organised ways to present information.

They arrange data clearly using **rows (across)** and **columns (down)** so you can read exact values quickly without guessing.

Unlike graphs, tables usually show **precise numbers rather than visual trends**, so accuracy matters.

Best for: Exact values and multi-step problems.

- ✓ *Scores.*
- ✓ *Distances.*
- ✓ *Totals.*
- ✓ *Schedules.*

Why Tables Are Useful

- → Every number is precise.
- → Great for totals and comparisons.
- → Often used in real-world data.

How to Read Tables Effectively

Before calculating anything:

1. Read the row and column headings carefully.
2. Check what each number represents.
3. Look for totals, patterns, or relationships.
4. Then answer the question.

Many mistakes happen because students rush this step.

Quick Memory Tips

Tables are like spreadsheets. Neat and organised.

Worked Example

Question: The average number of correct answers for all four students was **45**. **Nina** answered **6 more** questions correctly than Tom. How many correct answers did **Nina** get?

Student	Ali	Sara	Tom	Nina
Correct Answers	36	42	?	?

Steps:

Step 1 — Use the average to find the total

Total = Average × 4

Total = 45 × 4 = 180

So: **Ali + Sara + Tom + Nina**

$36 + 42 + x + y = 180$

$78 + x + y = 180$

$x + y = 102$ (Subtract 78 from both sides)

Step 2: Nina has 6 more than Tom:

$y = x + 6$

Step 3: So we have:

$x + y = 102$

$x + (\boldsymbol{x + 6}) = 102$ (In the **step 2** we have $y = x + 6$)

$2x + 6 = 102$

$2x = 96$

$x = 48$

So, **Tom** answered: **48**

Step 4: Find Nina's score

$y = \boldsymbol{x + 6} = 48 + 6 = \mathbf{54}$

✓ **Answer:** Nina got **54** correct answers.

Graph Summary Table

Graph Type	Best For	Key Question
Bar Graph	Comparing groups	Which is more/less?
Line Graph	Change over time	Is it increasing/decreasing?
Pie Chart	Parts of a whole	What fraction/percentage?
Table	Exact values	What is the precise number?
Pictograph	Visual counting	How many? (Check the key!)

The Scale Trap

Most students look at the bar and guess the number.

Common Traps

- ☒ The scale might jump by **2s, 5s, 10s, or 100s**
- ☒ The axis may not start at **zero**
- ☒ Pictographs may use symbols worth more than **1**

Common Question Types

Graph questions often ask the same things:

- ✓ How many more were sold on Friday than Monday?
- ✓ Which month had the greatest increase?
- ✓ What percentage is the largest slice?
- ✓ What is the total?
- ✓ What is the missing value?

If you read graphs like stories, these become quick wins.

The FAST CHECK Method (5 Seconds)

Before doing any maths, always do this:

Read the title — What is the graph about?

Check the key/legend — What do colours or symbols mean?

Check the scale — Counting by 1s, 2s, 5s, or 10s?

✗ Don't calculate yet.
✓ Now read the question.

This prevents careless mistakes.

Shortcuts for Success

<table>
<tr><th>Compare, Don't Calculate</th><th>The Pencil Trace</th></tr>
<tr><td>
Multiple-Choice Shortcut
When asked about:
✓ Biggest value.
✓ Smallest value.
✓ Greatest increase.
✓ Largest drop.
Do this:
→ Compare visually first
→ Eliminate wrong answers
→ Only calculate if needed
Many questions become 5-second wins.
</td><td>
The Pencil Trace Technique
When asked: "What is the value?"
1. Place your pencil at the top of the bar or point.
2. Trace across to the axis.
3. Read the number exactly.
Your eyes guess. Your pencil doesn't.
</td></tr>
</table>

Axes and Trends

X-axis = across (horizontal)

Y-axis = up (vertical)

Mixing them up costs easy marks.

Reading Trends

What You See	Meaning
↗ Upward line	Increasing
↘ Downward line	Decreasing
→ Flat line	No change

Try It Yourself

1. What type of graph shows change over time?
   A) Bar B) Line C) Pie D) Table
2. A pie chart shows:
   Red = 40%, Blue = 30%, Green = 20%. What percentage is Yellow?
3. A bar graph shows:
   Dogs = 5, Cats = 3, Birds = 4. Which is most? How many more than cats?
4. A line graph rises from left to right. Is the value increasing or decreasing?

Key Takeaway

Bars compare, lines show change.
Pie shows parts, tables arrange.
Read the labels, check the scale,
Spot the trend — you will not fail!

You don't need to redraw everything. Scan first. Trace the line. Solve with confidence

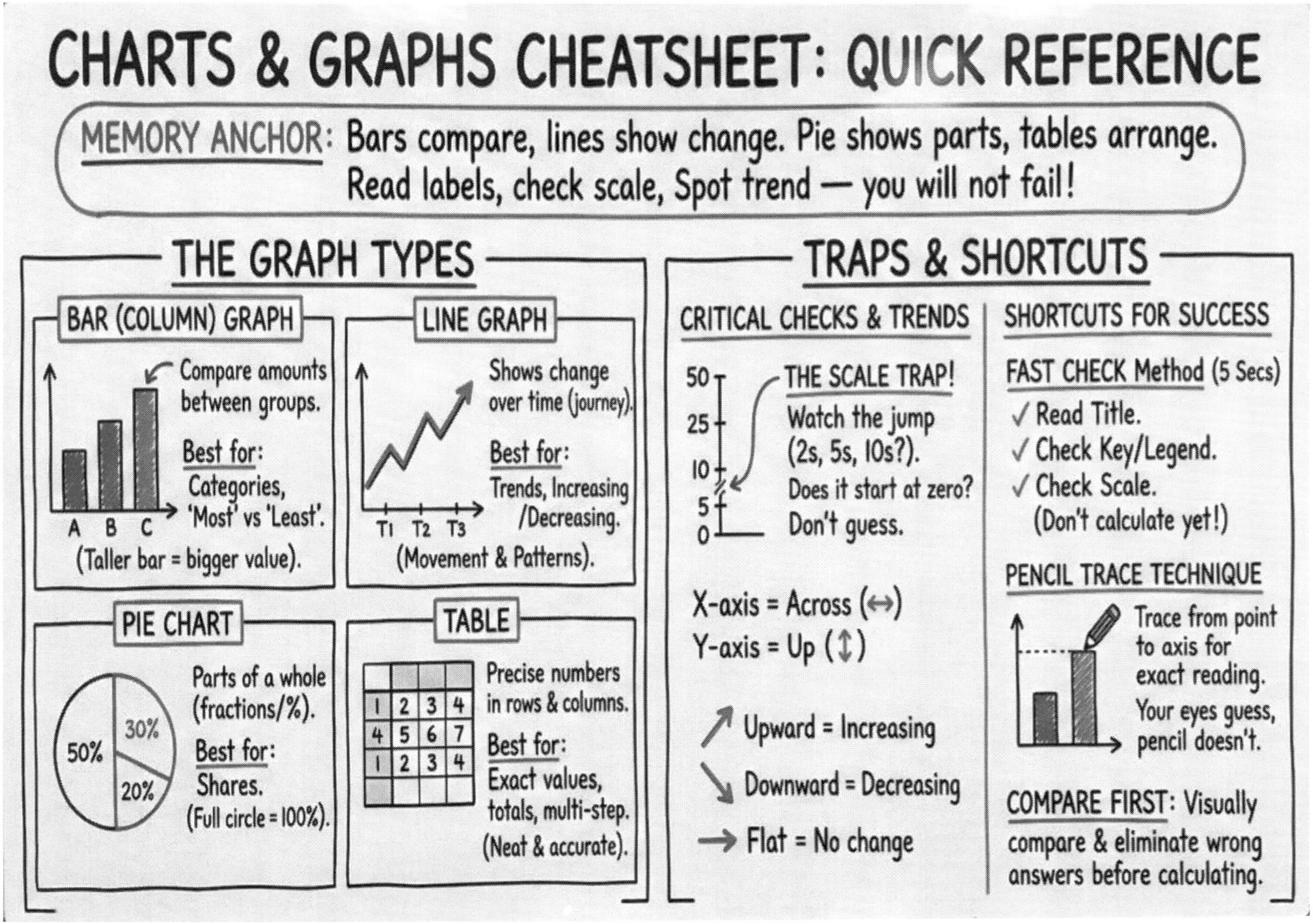
CHARTS & GRAPHS CHEATSHEET: QUICK REFERENCE
MEMORY ANCHOR: Bars compare, lines show change. Pie shows parts, tables arrange.
Read labels, check scale, Spot trend — you will not fail!
THE GRAPH TYPES
BAR (COLUMN) GRAPH
Compare amounts between groups.
Best for: Categories, 'Most' vs 'Least'.
A B C
(Taller bar = bigger value).
LINE GRAPH
Shows change over time (journey).
Best for: Trends, Increasing /Decreasing.
T1 T2 T3
(Movement & Patterns).
PIE CHART
50% 30% 20%
Parts of a whole (fractions/%).
Best for: Shares.
(Full circle = 100%).
TABLE
1 2 3 4
4 5 6 7
1 2 3 4
Precise numbers in rows & columns.
Best for: Exact values, totals, multi-step.
(Neat & accurate).
TRAPS & SHORTCUTS
CRITICAL CHECKS & TRENDS
50 25 10 5 0
THE SCALE TRAP!
Watch the jump (2s, 5s, 10s?).
Does it start at zero?
Don't guess.
X-axis = Across (↔)
Y-axis = Up (↕)
Upward = Increasing
Downward = Decreasing
Flat = No change
SHORTCUTS FOR SUCCESS
FAST CHECK Method (5 Secs)
✓ Read Title.
✓ Check Key/Legend.
✓ Check Scale.
(Don't calculate yet!)
PENCIL TRACE TECHNIQUE
Trace from point to axis for exact reading.
Your eyes guess, pencil doesn't.
COMPARE FIRST: Visually compare & eliminate wrong answers before calculating.

Probability & Chance

What Could Happen? How Likely Is It? — Predict Smart. Guess Less.

Some events are certain. Some might happen. Some will never happen.

Probability helps you:

- ✓ Eliminate impossible answers
- ✓ Avoid guessing
- ✓ Solve worded problems faster
- ✓ Make sense of fractions, decimals, and percentages

In OC and Selective School exams, probability is really about clear thinking, careful counting, and logical reasoning, not luck.

When you think in outcomes instead of confusion, questions become much easier.

Your Memory Anchor

Want over all,

one in total,

one for sure,

zero for not.

Probability made simple.

Meaning:

- ✓ **Want over all** → Probability = favourable ÷ total
- ✓ **One in total** → All probabilities together = 1
- ✓ **One for sure** → Certain = 1
- ✓ **Zero for not** → Impossible = 0

What Is Probability?

Probability measures how likely something is to happen.

It always sits between:

- ✓ **0 → Impossible** (cannot happen)
- ✓ **1 → Certain** (will happen)
- ✓ **Between 0 and 1 → Possible**

Probability can be written as:

- ✓ A fraction (most common in exams)
- ✓ A decimal
- ✓ A percentage

The Language of Chance

Understanding key words helps you interpret questions quickly.

Term	Meaning	Probability
Certain	Definitely happens	1
Impossible	Cannot happen	0
Likely	More chance than not	Between 0.5 and 1
Unlikely	Less chance than not	Between 0 and 0.5
Even chance	Equal yes/no chance	0.5

Understanding these words often lets you eliminate wrong answers fast.

Basic Probability

We are finding the chance of a particular outcome.

Formula

Probability = Favourable outcomes ÷ Total outcomes

Top = what we want
Bottom = everything possible

Always count carefully first.

Worked Example

Question: A bag has 3 red balls and 1 blue ball. What is the probability of picking a red ball?

Steps:

Favourable (want red) = 3
Total balls = 3 + 1 = 4

$$P(\text{red}) = \frac{3}{4}$$

Decimal form:

$$\frac{3}{4} = 0.75$$

✓ **Answer: 3/4 (or 0.75)**

Fraction form is usually preferred unless the question asks for decimals.

Total Probability = 1

In any probability situation, all possible outcomes together always add up to **1**, because something must happen. So, if you already know the probability of one event, you can find the probability of the opposite event by subtracting it from 1. This is called the **complement method**:

$$P(A) + P(\text{not A}) = 1$$

So, Probability of an event + Probability of it NOT happening = 1

We are finding the probability of the **opposite event**.

For example:

- ✓ Rain vs No rain
- ✓ Win vs Not win
- ✓ Red ball vs Not red ball

Worked Example

Question: If the probability of rain tomorrow is **0.3**, what is the probability that it will **not rain** tomorrow?

Steps:

$$\boldsymbol{P(\text{Rain}) + P(\text{not Rain}) = 1}$$

$$\boldsymbol{P(\text{not Rain}) = 1 - P(\text{Rain})}$$

$$\boldsymbol{P(\text{not Rain}) = 1 - 0.3 = 0.7}$$

✓ **Answer:** 0.7

Simple Events

A simple event involves just one action:

- ✓ One coin toss.
- ✓ One die roll.
- ✓ One marble draw.

Method Reminder:

1. Count all outcomes.
2. Count what you want.
3. Write favourable ÷ total.

Worked Example

Question: A spinner has A, B, C, D equally likely. What is the probability of landing on B?

Steps:

Favourable = 1, Total = 4

$\boldsymbol{P(B)}$ = Favourable ÷ total

$$= \frac{1}{4}$$

✓ **Answer:** $\frac{1}{4}$

"AND THEN" Probability

When two events happen one after another.

Clue words:

- ✓ and then
- ✓ followed by
- ✓ after that

.

💡 Chance that both events happen

$\boldsymbol{P(A\ and\ B)\ = P(A) \times P(B)}$

We multiply because both events must occur.

Worked Example

Question: A coin is tossed, then a die is rolled. What is the probability of heads and then a 6?

$\boldsymbol{P(\text{heads}) = \frac{1}{2}}$

$\boldsymbol{P(6) = \frac{1}{6}}$

Note:

Coin: A coin has 2 possible outcomes, heads or tails. **So,** $\frac{1}{2}$

Dice: A die has 6 possible outcomes, numbers 1 to 6. **So,** $\frac{1}{6}$

Steps:

Multiply:

$$\frac{1}{2} \times \frac{1}{6} = \frac{1}{12}$$

✓ **Answer:** $\frac{1}{12}$

These events do not affect each other, so they are independent.

Dependent Events (Without Replacement)

Sometimes the first choice changes the second choice.

This happens when the first item is **taken away and not returned**.

Examples:

- ✓ Pick a marble and keep it out of the bag
- ✓ Draw a card and keep it in your hand

Because one item is gone, the total number becomes smaller for the next step. So always update the numbers before calculating again.

Worked Example

Question: A bag has 3 red balls and 2 blue balls. One ball is picked and kept out. Then another ball is picked. What is the probability of picking a red ball first and then a blue ball?

Steps:

First pick red: $\frac{3}{5}$

After removing one red:

- Reds left = 2
- Blues left = 2
- Total left = 4

Second pick blue: $\frac{2}{4} = \frac{1}{2}$

Multiply: $\frac{3}{5} \times \frac{1}{2} = \frac{3}{10}$

✓ Final answer: $\frac{3}{10}$

"OR" Probability (Add Method)

Used when either event can happen but not both.

Example:

- ✓ Rolling a 2 or 5 on a die
- ✓ Picking a red or blue ball (single pick)

💡 Chance of either outcome

$P(A \ or \ B) \ = P(A + \ P(B)$

Worked Example

Question: What is the probability of rolling a 2 or a 5 on a fair six-sided die?

Steps:

$$\frac{1}{6} + \frac{1}{6} = \frac{2}{6} = \frac{1}{3}$$

Note: In both cases, there is 1 chance out of 6.

✓ Final answer: $\frac{1}{3}$

Fast Check Before Final Answer (5 Seconds)

Ask yourself:

- ✓ Is answer between 0 and 1?
- ✓ Did I count all outcomes?
- ✓ Did I simplify fractions?
- ✓ Does it make common sense?

Common Probability Traps

Trap	What Goes Wrong	How to Avoid
Forgetting total outcomes	Only counting favourable	Count everything first
Not simplifying	Writing 4/8 instead of 1/2	Simplify final fraction
Mixing AND/OR	Adding instead of multiplying	Check wording carefully
Ignoring replacement	Totals don't change when they should	Update numbers

Multiple-Choice Strategy

Before calculating:

- ✓ Cross out answers > 1
- ✓ Cross out negative values
- ✓ Impossible → answer = 0
- ✓ Certain → answer = 1

Eliminate nonsense first.

Probability Sense Guide

Event Type	Probability	Example
Impossible	0	Rolling a 10 on a standard die
Unlikely	$0 < P < 0.5$	Rolling a 6
Even chance	0.5	Heads on a fair coin
Likely	$0.5 < P < 1$	Drawing non-ace card
Certain	1	Rolling 1–6 on a die

Quick Checklist Summary

Step	Ask Yourself
1	What are all possible outcomes?
2	How many outcomes do I want?
3	Is it "not"? Use 1 – P
4	Is it "and then"? Multiply
5	Is it "or"? Add
6	Does the answer make sense?

Try It Yourself

1. A bag has 2 red and 3 blue balls. Find P(red).
2. If P(rain)=0.3, find P(no rain).
3. Classify as certain, likely, or impossible:
    - Sun rising tomorrow
    - Rolling a 10 on a die
    - Tossing heads on a coin
4. Toss a coin then roll a die. Find P(heads then 6).
5. Which is more likely?
   A) 1/10 B) 3/4 C) 0.05 D) 1
6. A jar has 5 red and 5 green beads. Two drawn without replacement. Find P(both green).
7. What is P(rolling a 2 or 4 on a die)?
8. A card is drawn from a deck. Find P(heart).

Key Takeaway

Want over all,

one in total,

one for sure,

zero for not.

Probability made simple.

You don't need complicated formulas. Think smart. Eliminate fast. Solve with confidence.

PROBABILITY & CHANCE CHEATSHEET: Predict Smart. Guess Less.

CORE CONCEPTS & ANCHOR

MEMORY ANCHOR: Want over all, one in total, one for sure, zero for not.

THE LANGUAGE OF CHANCE

CERTAIN (Probability = 1)

MIGHT HAPPEN ($0 < P < 1$)

IMPOSSIBLE (Probability = 0)

THE BASIC FORMULA

$$P(\text{event}) = \left[\begin{array}{c}\text{Favourable}\\\text{Outcomes}\\\text{(Want)}\end{array}\right] \div \left[\begin{array}{c}\text{Total}\\\text{Outcomes}\\\text{(All)}\end{array}\right]$$

→ As Fraction or Decimal.

TOTALS & TWO-STEP EVENTS

TOTAL PROBABILITY = 1

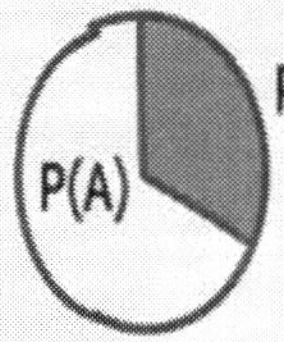

$$P(A) + P(\text{not } A) = 1$$

SIMPLE EVENTS (One Step)

1. Count ALL Outcomes → 2. Count WANT Outcomes → 3. DIVIDE (Favourable ÷ Total)

TWO-STEP EVENTS ("AND THEN...")

MULTIPLICATION RULE:

$$P(A \text{ and } B) = P(A) \times P(B)$$

INDEPENDENT: Events don't affect each other. Just multiply.

DEPENDENT (No Replacement): Second event depends on first.

CRITICAL: Update TOTAL numbers left *before* calculating second probability!

STRATEGY & SENSE CHECK

FAST CHECK (5 Seconds)

✓ Answer between 0 and 1?

✓ Do all parts add to 1?

✓ Common sense? (Certain/Likely/Impossible)

AVOID TRAPS & ELIMINATE

✗ Forgetting to count ALL outcomes first.

✗ Not simplifying fractions.

✗ Mixing up 'more likely' vs larger number.

PROBABILITY SENSE CHECK

0	0 to 0.5	0.5	0.5 to 1	1
(Impossible)	(Unlikely)	(Even Chance)	(Likely)	(Certain)

KEY TAKEAWAY: Think smart. Eliminate fast. Solve with confidence.

Packing, Tiling & Maximum Pieces

Fit Smart. Cut Clean. Get the Most.

Some shapes waste space. Some shapes fit perfectly. Some cuts give you more pieces than you expect.

If you understand **packing, tiling, and cutting**, you will:

- ✓ Solve faster.
- ✓ Avoid messy drawings.
- ✓ Eliminate wrong answers.
- ✓ Score maximum marks with less work.

This chapter is about thinking smart, not working long.

Your Memory Anchor

No gaps, no overlaps — tiling fits tight.
Across and down — multiply to get it right.
Each cut adds one when the lines all cross.
All cuts cross — use the formula, no loss.

Part 1: Packing (Fit the Most)

Packing is about fitting as many shapes as possible into a space. Sometimes they **fit perfectly**. Sometimes **small gaps remain**.

The goal isn't perfection, it's efficiency.

In exams, packing problems usually involve rectangles, squares, or boxes. Once you recognise how shapes line up in rows and columns, the maths becomes simple and fast.

Method (Across × Down Method)

When shapes line up neatly:

1. Divide the big length by the small shape's length → how many fit **across**
2. Divide the big height by the small shape's height → how many fit **down**
3. Multiply across × down

This works when shapes form clear rows and columns.

Worked Example

Question: How many 2 cm × 2 cm squares fit inside a 6 cm × 4 cm rectangle?

Steps:

Step 1: First, count how many squares fit across the rectangle:

$\mathbf{6 \div 2 = 3}$

So, **3** squares fit across.

Step 2: Next, count how many squares fit down the rectangle:

$\mathbf{4 \div 2 = 2}$

So, **2** squares fit down.

Step 3: Now multiply:

$\mathbf{3 \times 2 = 6}$

✓ **Answer:** 6

Always Check Rotation

When packing rectangular shapes, the orientation matters.

A small rectangle can often be placed in two ways:

- ✓ Length across, width down
- ✓ Width across, length down (rotated 90°)

This simple rotation can sometimes allow **more shapes to fit**.

Why Rotation Changes the Result

Packing depends on how many shapes fit:

- ✓ Across the length
- ✓ Down the width

Rotating swaps those measurements.

Sometimes one direction divides more neatly than the other, which means:

- ✓ Less wasted space.
- ✓ More complete shapes.
- ✓ A higher total number.

That's why exam questions often expect you to test both orientations.

When Rotation Helps Most

You should check rotation when:

- ✓ The small shape is a rectangle.
- ✓ The question asks for the *maximum number.*
- ✓ The first calculation leaves unused space.

You usually don't need to rotate:

- Squares (rotation doesn't change them)
- Angle tiling questions
- Maximum pieces (cutting) problems

Think About It

- ✓ *Can I rotate or rearrange the shapes to fit more?*
- ✓ *Can I make tighter rows to reduce wasted space?*

Worked Example

Question: How many **5 cm × 2 cm rectangles** fit inside a **10 cm × 6 cm rectangle?**

Steps:

First orientation:

Across:

$10 \div 5 = 2$

Down:

$6 \div 2 = 3$

Total:

$2 \times 3 = 6$

Rotate the rectangle:

Across:

$10 \div 2 = 5$

Down:

$6 \div 5 = 1$

Total:

$5 \times 1 = 5$

So, the original orientation fits more.

✓ **Answer:** Best fit = **6 rectangles**

Part 2: Tiling (Perfect Coverage)

Tiling is about covering a surface perfectly.
No gaps. No overlaps.

Think of bathroom tiles or paving stones, every piece must fit exactly.

These questions often test whether shapes fit together neatly or whether their angles allow complete coverage.

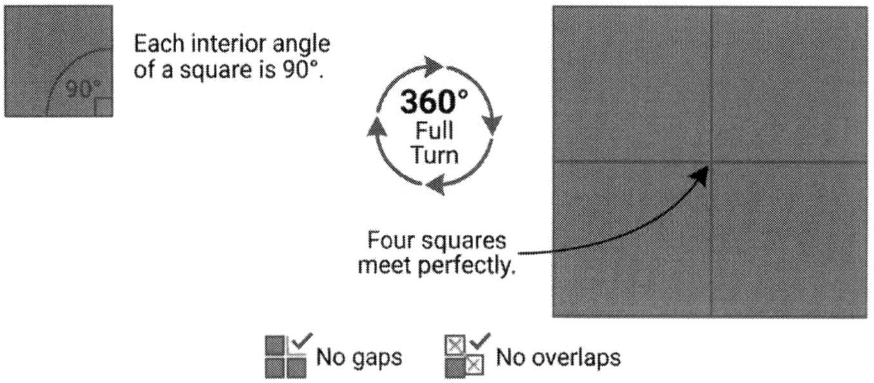

Two Common Types of Tiling Questions

1. Angle Check (Shapes Meeting at a Point)

When regular shapes meet at a point:

360° ÷ interior angle

- ✓ Whole number → shape tiles
- ✓ Not whole → gaps or overlaps

This rule applies when **regular shapes tile by themselves**.

Square Tiles

Interior angle = 90°

360° ÷ 90° = 4

Four squares meet perfectly.

✔ **Answer:** Tiles: The shape fits together perfectly to cover a surface with no gaps or overlaps.

Regular Pentagon

Interior angle = 108°

360° ÷ 108° ≈ 3.33

Not whole:

- 3 pentagons → **gap**
- 4 pentagons → **overlap**

✓ **Answer:** Does not tile. The shape cannot cover a surface perfectly because it leaves gaps or causes overlaps.

Regular Shapes That Tile

- Equilateral triangle
- Square
- Rectangle
- Regular hexagon

Regular Shapes That Do NOT Tile Alone

- Regular pentagon
- Heptagon
- Octagon
- Circle (always leaves gaps)

2. Tiling Inside Rectangles

Here we use measurements rather than angles.

Perfect tiling requires:

- Exact fit across
- Exact fit down
- Whole-number divisions

Method

1. Divide length
2. Divide width
3. BOTH must be whole numbers
4. Multiply

If either division isn't whole, it is not perfect tiling.

Worked Example

Question: How many 2 cm squares fit inside a 10 cm × 6 cm rectangle?

Steps:

Across:

$10 \div 2 = 5$

Down:

$6 \div 2 = 3$

Multiply:

$5 \times 3 = 15$

✓ **Answer:** 15 tiles

Worked Example

Question: Can 3 cm squares tile a 10 cm × 6 cm rectangle?

Steps:

Across: $10 \div 3 \approx 3.33$ ✗
Down: $6 \div 3 = 2$

Not whole in both directions.

✓ **Answer:** ✗ Not perfect tiling.

Tiling Checklist

- ✓ Did I divide both directions?
- ✓ Are BOTH whole numbers?
- ✓ Any gaps?
- ✓ Any overlaps?

> ⚠ **Important**
>
> Both divisions must be whole numbers. If not, gaps appear and perfect tiling is impossible.

Part 3: Maximum Pieces (Cutting Problems)

These problems ask how many pieces you can get from cuts, but drawing every cut quickly becomes messy.

Instead, we look for patterns.

When each new straight cut crosses all previous cuts, the number of pieces increases rapidly. Using a simple formula makes these questions quick and accurate.

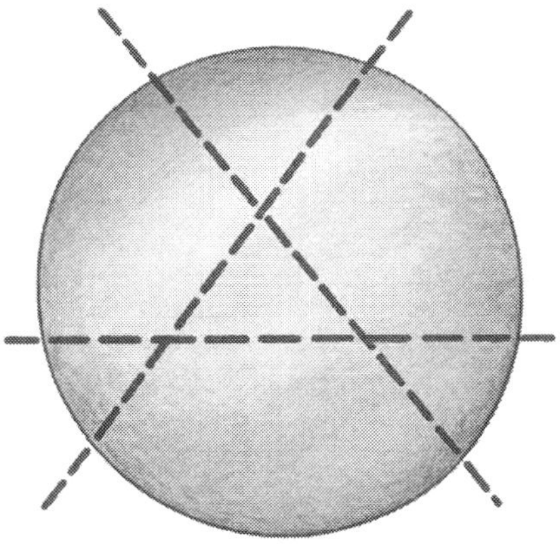

Pattern First (Understand Before Formula)

Cuts (n)	**1**	**2**	**3**	**4**	**5**	**6**	**7**	**8**	**10**
Max Pieces $P(n)$	2	4	7	11	16	22	29	37	56

Notice the pattern: Each new cut adds one more piece than the previous increase (+1, +2, +3, +4...). This happens because every new cut crosses all earlier cuts, creating extra new regions.

The Formula

$$P(n) = \frac{n^2+n+2}{2}$$

Where:

- n = number of cuts
- P(n) = maximum number of pieces

This works when:

✓ Cuts are straight.

✓ Each cut crosses all others.

Otherwise, fewer pieces form.

Worked Example

Maximum pieces from 3 cuts.

$$P(3) = \frac{3^2+3+2}{2}$$

$$= \frac{9+3+2}{2}$$

$$= \frac{14}{2} = 7$$

✓ **Answer:** 7 pieces, the maximum number possible when 3 straight cuts all cross each other.

Common Mistakes

✓ Curved cuts → formula doesn't apply.

✓ Parallel cuts → fewer pieces.

✓ Cuts not crossing → fewer pieces.

Always check the conditions.

Quick Summary Table

Topic	**Strategy**	**Key Rule**
Packing	Divide across & down	Multiply totals
Tiling (Rectangle)	Divide both directions	BOTH whole numbers
Tiling (Angles)	360° ÷ angle	Whole number needed
Maximum Pieces	Use formula/pattern	Cuts must cross

What These Questions Look Like

You might see:

- ✓ Which shape can tile a floor?
- ✓ How many small squares fit inside this rectangle?
- ✓ What is the maximum number of pieces after 7 straight cuts?
- ✓ Which arrangement wastes the least space?

With shortcuts, these become easy wins.

Try It Yourself

1. Which shape can tile a floor?
   A) Circle B) Regular hexagon C) Pentagon D) Star
2. How many 5 cm squares fit inside a 20 cm × 10 cm rectangle?
3. Can 4 cm × 3 cm rectangles tile a 12 cm × 8 cm rectangle? Why?
4. What is the maximum number of pieces from 7 straight cuts?
5. A floor is 15 m by 6 m. Tiles are 3 m by 2 m. How many tiles are needed?

Key Takeaway

No gaps, no overlaps — tiling fits tight.
Across and down — multiply to get it right.
Each cut adds one when the lines all cross.
All cuts cross — use the formula, no loss.

You don't guess. You divide. You rotate. You use formulas. You solve fast.
Think smart. Enjoy maths. Win time in the test

PACKING, TILING & MAXIMUM PIECES: SMART STRATEGIES

PACKING (Fitting Smart)

Goal: Fit most shapes, small gaps allowed.

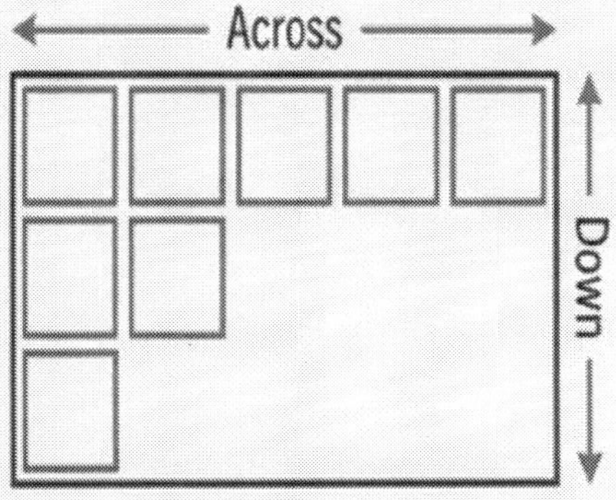

Calculation:

(Big Length ÷ Small Length)
× (Big Height ÷ Small Height)
= Total Shapes

Think About It:

- Rotate shapes to fit more?
- Tighter rows = less waste

TILING (No Gaps, No Overlaps)

Goal: Perfect coverage.

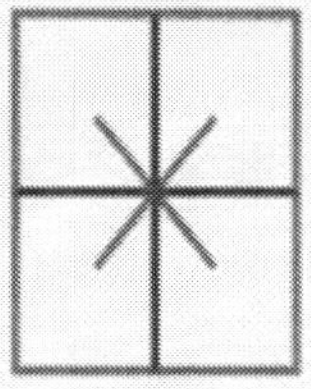

Angle Check:
360° ÷ Angle
= Whole Number

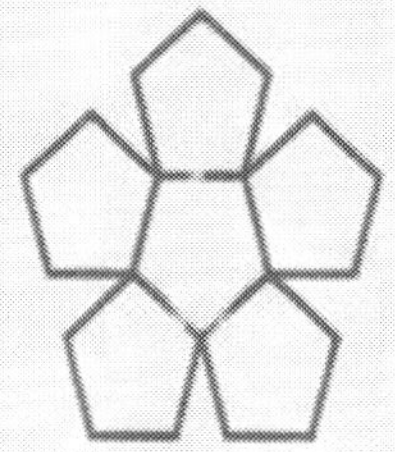

Not a whole number
= Gaps

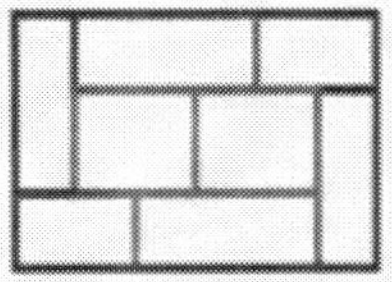

Length Check (Rectangles):
Big Dim. ÷ Small Dim.
= Whole Number
(Both ways)

Common Shapes:

Tile:		Don't Tile:	
Square	✓	Circle	✗
Rectangle	✓	Regular Pentagon	✗
Equilateral Triangle	✓	Star	✗
Regular Hexagon	✓		

MAXIMUM PIECES (Cutting Clean)

Goal: Greatest pieces from straight cuts.

Formula (Lazy Caterer's):

$$P(n) = \frac{n^2 + n + 2}{2}$$

Where n = cuts, P(n) = max pieces

Quick Reference Table

Cuts (n)	Max Pieces P(n)
1	2
2	4
3	7
4	11
5	16
6	22
7	29

When Formula Works:

✓ Straight cuts
✓ Each new cut crosses all previous
✓ Curved cuts
✗ Cuts don't cross

MEMORY ANCHOR | No gaps, no overlaps — tiling fits tight. Across and down — multiply to get it right. Each cut adds one when the lines all cross. All cuts cross — use the formula, no loss.

Rates, Change & Flow

Fill Fast. Leak Smart. Think in Net Speed.

Some things flow in. Some things leak out. Sometimes both happen at the same time, and that is when students often panic.

But here's the truth: These problems are simply rates working together. Not speed alone, but change over time.

When you understand the **flow personality** of a question:

- ✓ You solve faster
- ✓ Avoid messy tables
- ✓ Spot impossible answers quickly
- ✓ Stay calm in multi-step problems

This chapter turns tricky flow questions into manageable, logical steps, exactly what OC and Selective exams reward.

Your Memory Anchor

In minus Out — that's the rate,
More in? It fills. More out? It drains its fate.
Same in, same out? No rise, no fall —
What you start with stays the same, that's all.

The Core Idea: Net Rate

To find how something is changing:

Net Rate = Inflow − Outflow

That's it. Subtract what leaves from what enters.

Condition	Result	Meaning
Net > 0	Level rises	Filling
Net < 0	Level falls	Emptying
Net = 0	No change	Steady state

Important Concept Link

Flow problems are really just **rate problems where multiple rates act together**.

Whether it's:

- ✓ Water pipes
- ✓ Fuel consumption
- ✓ Juice being poured and drunk
- ✓ Machines producing and removing items

The maths stays the same: Always think net change.

Fast Check (5-Second Thinking)

Before calculating anything, pause and ask:

- ✓ What flows IN?
- ✓ What flows OUT?
- ✓ Which one is faster?

This quick check often predicts the answer direction before any calculation. That's powerful exam strategy.

Two Types of Flow Problems

Type 1: Filling vs Leaking (Tanks, Pipes, Containers)

Something goes in while something leaks out.

Examples:

- → A pipe fills a tank while another pipe leaks
- → A bath runs while the plug is open

Ask yourself

Which is faster, filling or leaking?

Type 2: Pouring vs Consuming (Juice, Fuel, Water)

Something is added while something is being used.

Examples:

→ Fuel added while a car is running

→ Juice poured while people drink

Ask yourself

What is the net change each minute (or second)?

Time to Fill / Time to Empty

If:

✓ **Fill rate =** A units per minute

✓ **Leak rate =** B units per minute

Then:

Net Rate = A – B

To find time:

Time = Total amount ÷ Net rate

📌 **Simple rule**

Fill faster? It fills.
Leak faster? It drains.

Worked Example

Question: A pipe fills a tank in 10 minutes. A leak empties it in 20 minutes. How long will it take to fill the tank if both are open?

Think smart:

Fill rate = $\frac{1}{10}$ tank per minute

Leak rate = $\frac{1}{20}$ tank per minute

Net rate: Inflow – Outflow

$$\frac{1}{10} - \frac{1}{20} = \frac{2}{20} - \frac{1}{20} = \frac{1}{20} \text{ tank per minute}$$

Time: Total amount ÷ Net rate

$$1 \div \frac{1}{20} = 1 \times \frac{20}{1} = 20 \text{ minutes}$$

✓ **Answer:** 20 minutes

"Will It Ever Fill?" Logic

Some questions try to trick you. Use this simple check.

Condition	**Outcome**
Inflow > Outflow	Yes — it will fill
Inflow < Outflow	No — it won't fill
Inflow = Outflow	Level stays the same

Steady-State Situations

When inflow equals outflow, nothing changes. This is called a steady state.

You may see phrases like:

- ✓ Water level stays the same.
- ✓ Fuel amount does not change.
- ✓ Height remains constant.

📌 **Simple rule**

No change means Inflow = Outflow

Worked Example

Question: A tank **fills** at **4 L/min** and leaks at **4 L/min**. Will it ever fill?

Think smart:

Net rate = Inflow – Outflow = 4 – 4 = 0

No increase → no filling.

✓ **Answer:** It will never fill.

Question: A tank stays at the same level. One pipe adds 6 L/min. What must the leak rate be?

Think smart:

Since nothing changes:

Leak rate = 6 L/min.

✓ **Answer:** 6 L/min

Multi-Step Flow Problems

These look scary, but they're just several small steps done one at a time.

Strategy for Multi-Step Problems

1. Handle one time period at a time.
2. Find the net change for that period.
3. Update the amount.
4. Move to the next step.

Never mix different time steps together.

Worked Example

Question: A tank has **40 L** of water.

First 5 minutes: inflow 6 L/min, leak 2 L/min

Next 5 minutes: only a leak of 3 L/min

How much water remains?

Steps:

Step 1: First 5 minutes

Net rate = Inflow − Outflow **=** 6 − 2 = 4 L/min

Change = how much the amount increases or decreases over time.

= Net rate × Time
= 4 × 5 = 20 L

New amount:

= Old amount + Change

= 40 + 20 = 60 L

Step 2: Next 5 minutes

Net rate = −3 L/min
Change = −3 × 5 = -15L

Final amount:

60 − 15 = **45 L**

✓ **Answer:** 45 L

Common Traps to Avoid

Watch out for these:

- ☒ Adding rates instead of subtracting → Always do In – Out
- ☒ Forgetting leaks reduce the total → Outflow is negative change
- ☒ Mixing time periods → Solve one step at a time
- ☒ Assuming filling just because water enters → Check if the leak is faster

Try It Yourself

1. A tap fills a tank in 12 minutes. A leak empties it in 18 minutes.
   How long will it take to fill the tank?
2. A bottle fills at 5 mL/s while 3 mL/s is used.
   What is the net change per second?
3. A container fills at the same rate it leaks.
   What happens to the water level?
4. A tank has 30 L:
5. 4 minutes: +5 L/min in, −1 L/min out
6. Next 2 minutes: only −3 L/min out
   How much water remains?
7. Will this tank ever fill?
   Fill = 2 L/min, Leak = 3 L/min.

Key Takeaway

In minus Out — that's the rate,
More in? It fills. More out? It drains its fate.
Same in, same out? No rise, no fall —
What you start with stays the same, that's all.

Don't guess. Don't overwork.

Think in net speed, and flow problems become easy.

RATES, CHANGE & FLOW: Fill Fast. Leak Smart. Think in Net Speed.

THE FOUNDATION: NET RATE

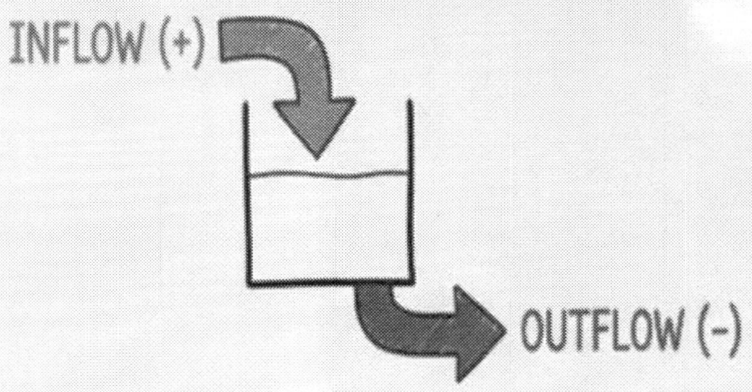

NET RATE = INFLOW − OUTFLOW

Memory Anchor:
In minus Out — that's the rate.
More in? It fills. More out? It drains its fate.
Same in, same out? No rise, no fall.

The 3 States

Net > 0	→ Level RISES (Filling) ⬆
Net < 0	→ Level FALLS (Emptying) ⬇
Net = 0	→ NO CHANGE (Steady State) —

QUICK LOGIC & FORMULAS

5-Second Fast Check

- ✓ What flows IN?
- ✓ What flows OUT?
- ✓ Which is faster?

Will It Ever Fill? (Simple Rule)

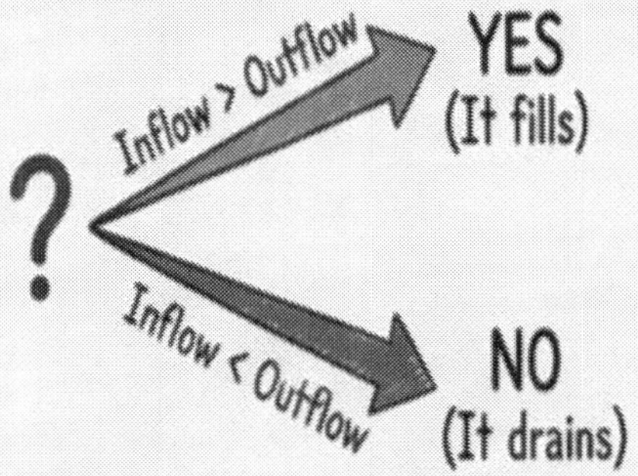

TIME = Total Amount ÷ Net Rate

CHANGE = Net Rate × Time

STRATEGY & WATCH-OUTS

Multi-Step Strategy

1. Handle ONE time period.
↓
2. Find NET change for period.
↓
3. Update total amount.
↓
4. Move to next step. (Never mix periods!)

Common Traps

- ✗ Adding rates instead of subtracting.
- ✗ Forgetting leaks reduce total (negative).
- ✗ Mixing time steps together.

KEY TAKEAWAY: Don't guess. Think in NET SPEED.

Coordinates, Maps & Location

Find It Fast. Move It Smart. Never Get Lost.

Some questions ask you to **read** a position. Some ask you to move a **position**. Some ask you to **measure** between positions. If you understand how location works, you'll answer faster, avoid careless direction mistakes, eliminate wrong options, and visualise movement instead of guessing. This chapter turns maps and coordinates into easy wins.

Your Memory Anchor

Flat then up, X then Y,

Side to side, then to the sky.

North is up, East to the right,

Follow the grid, you'll get it right.

Latitude & Longitude

Think of the Earth like a giant invisible grid that helps people describe exact locations, just like coordinates on graph paper.

Latitude

Latitude measures position **north or south of the Equator**. Lines of latitude run **horizontally (side to side)** around the globe.

They never meet each other, and they tell you **how far up or down you are from the middle of the Earth**.

Example: Sydney is south of the Equator, so it has a southern latitude.

Longitude

Longitude measures position east or west of the Prime Meridian (Greenwich). Lines of longitude run vertically from pole to pole.

Unlike latitude lines, they meet at the poles. They tell you how far left or right you are around the Earth.

Reading Coordinates

The Order Rule

Coordinates are written as (x, y), meaning:

- ✓ Move across first (x-direction)
- ✓ Then move up or down (y-direction)

This is like walking along a street first, then climbing stairs.

Worked Example

Point (3, 5) means:

1. Move 3 **across** to the right.
2. Then 5 **up.**

You end at that exact grid point.

Selective Trap

Many students go up first because that feels natural, but that gives the wrong point. ***Always follow X then Y.***

Grid References

Maps often divide areas into squares so locations can be found quickly and accurately, without guessing. This grid system helps everyone describe the same location clearly.

Method

1. Start from the origin (or bottom-left corner).
2. Move across first.
3. Then move up.

This keeps everyone reading the map the same way.

Across *you go, then* ***up*** *you climb, grid reading every time.*

If a square is labelled (for example **C4**):

1. The letter tells the column (**across**).
2. The number tells the row (**up**).

This is why library shelves, chessboards, and maps all use similar systems.

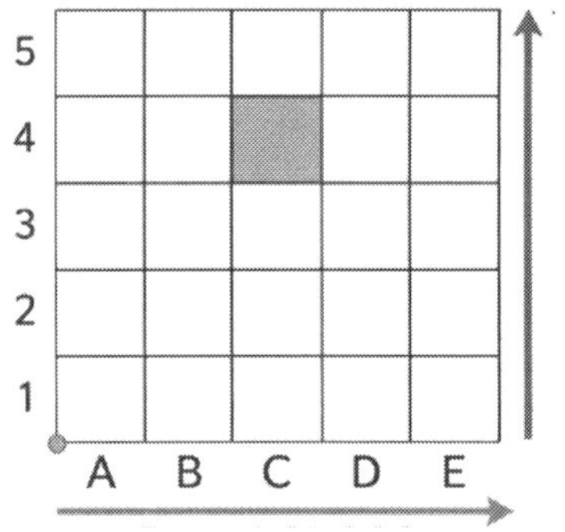

Distance Between Locations

On grid maps, distance is usually counted in squares rather than using a ruler.

Straight Moves

If movement is only horizontal or vertical:

→ Simply count the squares.

This works because each square represents equal distance.

Mixed Movement

If movement includes both directions, break it into two parts.

1. across movement.
2. up/down movement.
3. Then add them.

Distance $= | x_2 - x_1 | + | y_2 - y_1 |$

This is sometimes called grid distance or taxicab distance, like city streets where cars can't drive diagonally.

Worked Example

Grid Distance

A = (2, 1), B = (6, 4)

Across: $6 - 2 = 4$
Up: $4 - 1 = 3$

Total distance = 4 + 3 = **7 units.**

Side to side, then up you see, add the moves, that's the key.

Directions & Bearings

Maps always assume the same orientation unless told otherwise.

Direction positions

- North = up (↑)
- East = right (→)
- South = down (↓)
- West = left (←)

This matches coordinate grids:

- North → positive **y direction**
- East → positive **x direction**

So maps and maths usually align nicely.

Bearings

A **bearing** describes direction using an angle in degrees (°).

Instead of saying:

→ Go east

→ Head south-west

we give an exact direction like **045°**, **120°**, or **270°**.

Numbers make direction precise, no guessing.

Key Rule

Always measure clockwise from North.

Think of facing North first, then turning clockwise like a clock.

Direction	Bearing
North	000°
East	090°
South	180°
West	270°

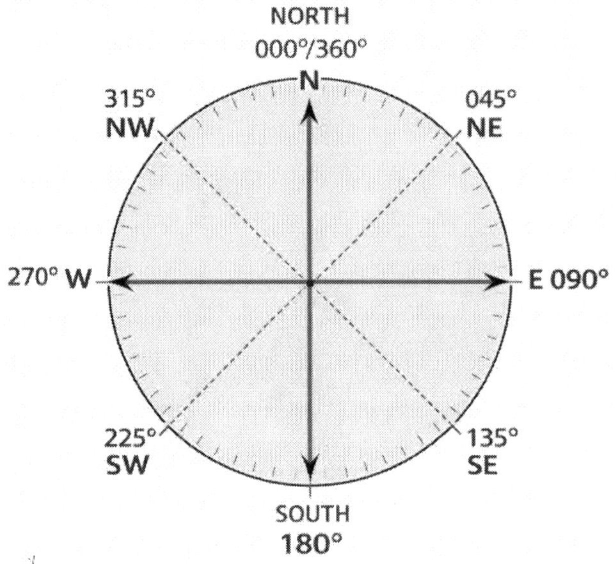

Bearings are written using **three digits** (e.g. **045°,** not 45°).

Why Bearings Are Useful?

Clear navigation is essential in many real-world situations. Pilots, ship captains, hikers, and GPS systems all rely on bearings to travel accurately and stay on course. Bearings are also commonly used in maps and exam questions, especially in selective tests, where they help assess a student's understanding of direction, map-reading ability, and logical spatial thinking.

When Bearings Are Used in Questions

Usually when:

- A compass or map is shown
- Direction must be exact

- Movement between locations is described
- Reverse direction is needed

Look for words like bearing, clockwise, compass, direction, north, east, south, west.

***North** is up, **East** is right, **South** goes down, **West** takes flight.*

Worked Example

Question: A plane travels north-east. What is its bearing?

Think smart:

1. Start facing **North (000°)**.
2. Turn clockwise toward **East (090°)**.
3. **North-east** is **halfway** between them.

$$(0° + 90°) \div \mathbf{2} = \mathbf{45°}$$

Write with three digits:

✓ **Answer: 045°**

Question: The bearing of town B from town A is **065°**. Find the bearing of town A from town B.

Steps:

1. Understand the wording.

- "B from A" → stand at A, face B.
- "A from B" → stand at B, look back at A.

You are facing along the same line but in the **opposite direction**.

2. Opposite directions differ by 180°.

Why?

- Full turn = 360°
- Half turn = 180°
- **Looking back = half turn.**

So: $\mathbf{065° + 180° = 245°}$

✓ **Answer:** Bearing of A from B = **245°**

Position After Movement

These questions test whether you can follow directions carefully and update positions correctly. Accuracy matters more than speed. Most mistakes happen when steps are rushed.

Movement Method

1. Start at the given coordinates.
2. Follow one direction at a time.
3. Update the coordinates after each move.

Do not combine steps too early. That often leads to errors

- ✓ One move. One update.

How Movement Changes Coordinates

Movement	What Happens
East (right)	x increases
West (left)	x decreases
North (up)	y increases
South (down)	y decreases

Quick Way to Think About It

- ✓ Horizontal movement = change in x (Across changes **x**).
- ✓ Vertical movement = change in y (Up or down changes **y**).

Common Student Trap

Students often:

- ☒ Change both x and y at once.
- ☒ Forget direction signs (+ / −).
- ☒ Skip intermediate steps.

Writing each step avoids these mistakes.

Worked Example

Question: Point P starts at **(−1, 3)**.

It moves:

- 4 units east
- 2 units south
- 3 units west

What are the final coordinates of P? See the diagram on the next page.

Steps:

Start: $(-1,3)$

Move 4 east → **x increases**

$$-1 + 4 = \mathbf{3}$$

New position: $(\mathbf{3}, 3)$

Move 2 south → **y decreases**

$$3 - 2 = \mathbf{1}$$

New position: $(3, \mathbf{1})$

Move 3 west → **x decreases**

$$3 - 3 = \mathbf{0}$$

Final position: $(\mathbf{0}, 1)$

✓ **Answer:** $(0, 1)$

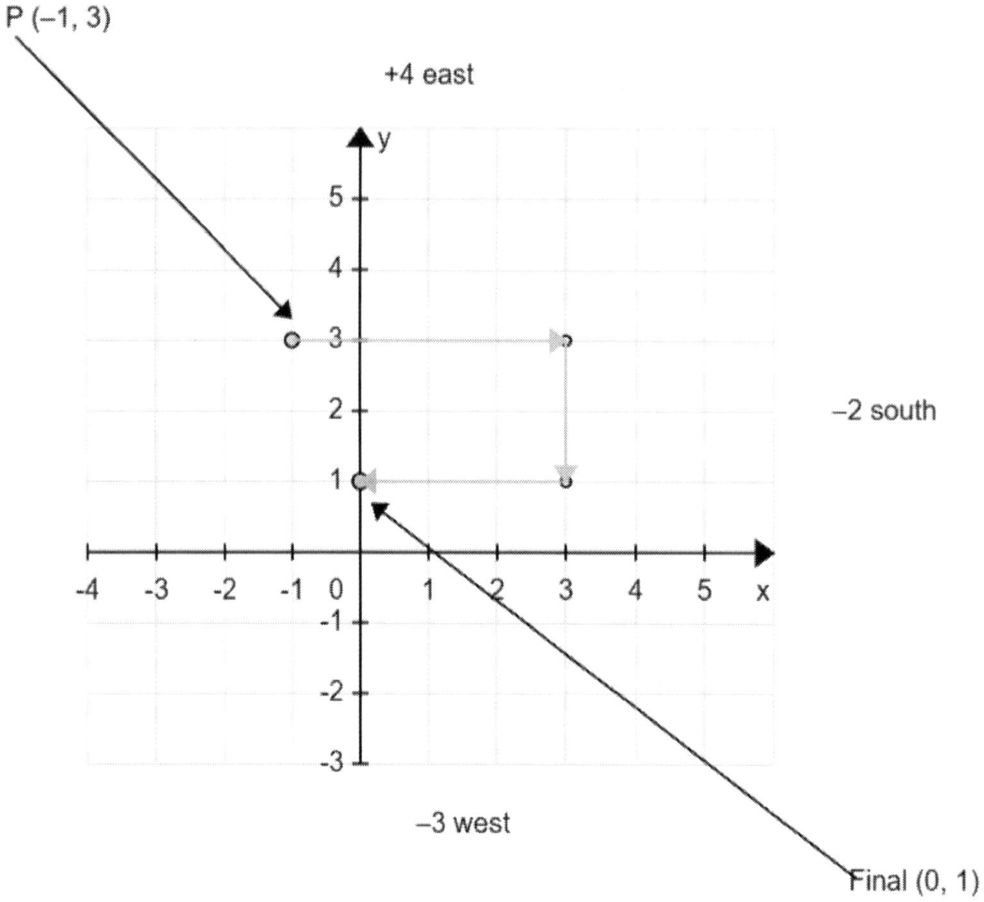

Common Question Types

You might see questions like:

- ✓ What are the coordinates of point A?
- ✓ Which point is at (–2, 4)?
- ✓ How far is A from B?
- ✓ After moving 3 east and 2 north, where do you land?
- ✓ Which location is north-west of point P?
- ✓ What is the bearing from A to B?

These look different, but they all use the same basic rules.

Once you master the system, these become fast marks.

Why This Chapter Matters

Coordinates and maps appear in:

- ✓ Geometry problems
- ✓ Word problems about movement
- ✓ Navigation and bearings
- ✓ Scale drawings
- ✓ Patterns and logic puzzles
- ✓ Real-world location problems

These skills connect maths to real life. Maps, GPS, gaming, travel, and science all use them.

Master it once, benefit everywhere.

Try It Yourself

1. What are the coordinates of a point 3 right and 5 up from the origin?
2. Starting at (4, 2), move 2 west and 3 north. Where do you land?
3. What is the distance between (1, 4) and (6, 4)?
4. Which direction is south-west?
5. A point moves: up 2, right 3, down 1. What is the net movement?
6. Which point is in the north-east quadrant?
   A) (−3, 4) B) (2, −5) C) (4, 3) D) (−2, −1)

Key Takeaway

Flat then up, X then Y,

Side to side, then to the sky.

North is up, East to the right,

Follow the grid, you'll get it right.

Location questions aren't about guessing. They're about following a system.

When you read carefully and move step by step, maps become simple.

COORDINATES, MAPS & LOCATION: QUICK REFERENCE CHEATSHEET

Memory Anchor: Flat then up, X then Y. Side to side, then to the sky.

LATITUDE & LONGITUDE

(The Earth Grid)

LATITUDE: North/South of Equator. Runs horizontally.

LONGITUDE: East/West of Prime Meridian. Runs pole-to-pole.

READING COORDINATES

(The (x, y) Order Rule)

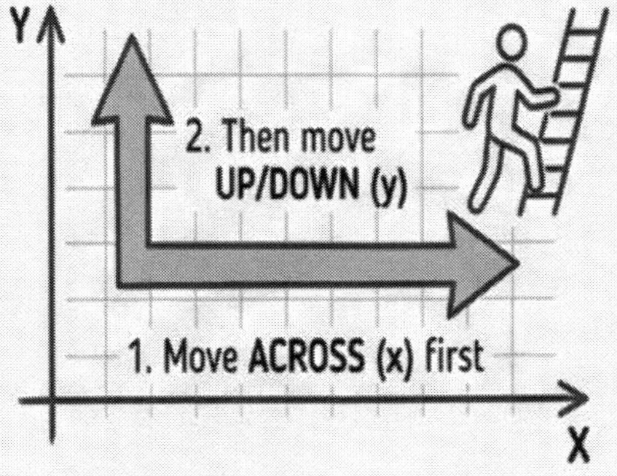

(x, y) means: Across then Up.

Avoid Trap: Never go "up" first.

GRID REFERENCES & DISTANCE

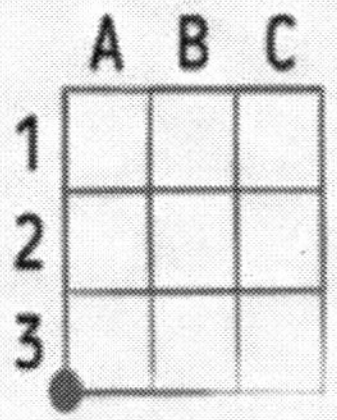

Method: Start at origin (bottom-left). Go Across, then Up. (e.g., C4 = Col C, Row 4).

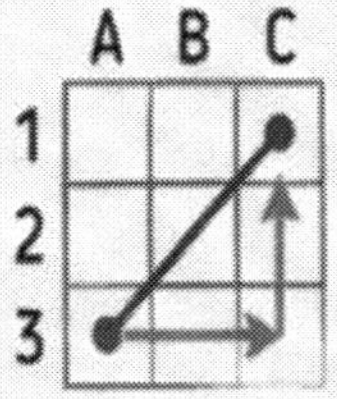

Calculating Distance
Straight moves: Count squares.
Mixed moves: Add the parts (Across distance + Up/Down distance).

POSITION AFTER MOVEMENT

(Updating Coordinates)

NORTH: y increases (+y)

WEST: x decreases (-x) ← (x,y) → **EAST:** x increases (+x)

SOUTH: y decreases (-y)

Rule: One direction at a time. Update step-by-step.

STANDARD DIRECTIONS

Matches coordinate grids (North = +y, East = +x).

BEARINGS (Precise Navigation)

Definition: Direction as an angle (°).
Always 3 digits (e.g., 045°).

MEASURE CLOCKWISE FROM NORTH.

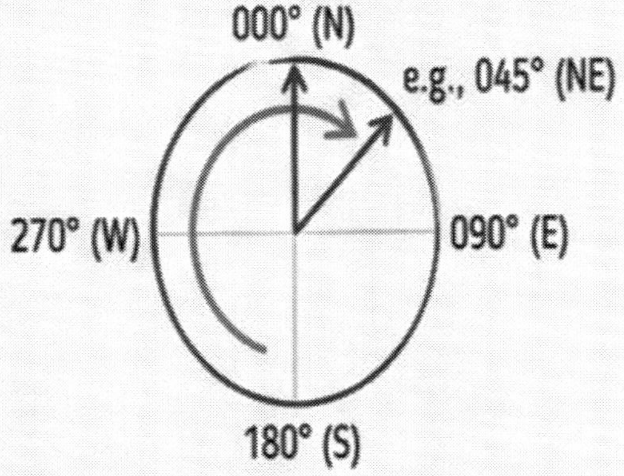

Back Bearings (Reverse Direction)

⇄ Opposite directions differ by 180°. (Add or subtract 180°).

Counting Methods & Arrangements

Spot the Pattern. Count Smarter. Score Faster.

Counting questions can look intimidating at first, with lots of options, lots of "How many ways?", and lots of chances to make small mistakes. But most counting problems follow a few patterns. Once you recognise them, these questions often become quick, reliable marks.

The goal is to **spot the structure** and use the right shortcut.

Your Memory Anchor

Choose, then choose? Multiply the clues.
Same thing twice? Divide it nice.
Grid moves right and down? Count paths, don't drown.

The Fundamental Counting Principle

Most counting problems reduce to three core ideas:

- ✓ **Choices** (outfits, menus, selections)
- ✓ **Paths** (movement on grids)
- ✓ **Arrangements** (ordering things)

Recognising which type you have often makes the problem much easier.

Multiplication Principle

If you make one choice and then another, and the second choice does not depend on the first (they are **independent**), multiply the number of options at each step.

Choose → Choose → Multiply.

Independent means:

- ✓ Choosing one option does not remove other options.
- ✓ Every first choice can pair with every second choice.

Worked Example

Ice Cream Choices

An ice-cream shop offers:

- 2 flavours (chocolate, vanilla)
- 2 cone types (waffle cone, cup)

To order ice cream:

1. Choose a flavour.
2. Choose a cone.

Each flavour works with any cone:

1. Chocolate + waffle.
2. Chocolate + cup.
3. Vanilla + waffle.
4. Vanilla + cup.

So, $\mathbf{2 \times 2 = 4}$ **choices**

This shows why multiplication works. Every first choice pairs with every second choice.

The Category Choice Rule

If you choose one item from each of $k$separate categories, and category $i$has $n_i$options:

$$\text{Total outcomes} = \boldsymbol{n_1} \times \boldsymbol{n_2} \times \cdots \times \boldsymbol{n_k}$$

This is a direct application of the multiplication principle.

Important Note

Although this is often introduced using meals, the rule is **not just about food**. It applies whenever you choose one item from each category.

Examples:

- ✓ Meals (starter + main + dessert)
- ✓ Outfits (shirt + pants + shoes)
- ✓ Ice cream (flavour + cone)
- ✓ Passwords (letter + number)
- ✓ Subject choices at school

Worked Example

Menu Choices

A restaurant offers:

- 2 starters
- 4 mains
- 3 desserts

Number of meals:

$$\mathbf{2 \times 4 \times 3 = 24}$$

Multiplication avoids listing every possibility.

Counting with Restrictions (Exclusion Method)

Sometimes there are extra conditions, such as:

- "At least one must be..."
- "Cannot choose both..."
- "Not allowed..."

Direct counting can be complicated. Instead, we use the **exclusion method**.

General Observation

Often it is easier to:

✓ Count all possibilities first, then
✓ Subtract unwanted cases.

$$\textbf{Valid outcomes} = \textbf{Total outcomes} - \textbf{Unwanted outcomes}$$

Worked Example

Question: A student has:

- 4 T-shirts (2 plain, 2 striped)
- 3 shorts (1 black, 2 coloured)

Find the number of outfits if **at least one item must be plain or black.**

Steps:

Step 1: Count all outfits:

$$4 \times 3 = \mathbf{12}$$

Step 2: Count unwanted outfits. Striped shirt AND coloured shorts:

$$2 \times 2 = \mathbf{4}$$

Step 3: Subtract:

$$12 - 4 = \mathbf{8}$$

✓ **Answer:** So, there are **8 valid outfits.**

Arrangements & Factorials

Sometimes **order matters**. This leads to factorials.

- ABC is different from BAC.
- First place is different from second place.

Factorial

For a positive integer n:

$$\boldsymbol{n! = n \times (n-1) \times (n-2) \times \cdots \times 1}$$

A factorial counts how many ways we can arrange $\boldsymbol{n}$ distinct items in order.

This comes from repeatedly applying the multiplication principle:

- $\boldsymbol{n}$ choices for the first position
- $\boldsymbol{n-1}$ choices for the next
- $\boldsymbol{n-2}$ for the next
- continuing down to $\mathbf{1}$

Worked Example

$$3! = 3 \times (3-1) \times (3-2)$$
$$= 3 \times 2 \times 1$$
$$= 6$$

Factorial Quick Reference Table

n	n!	Calculation
1	1	1
2	2	2×1
3	6	$3 \times 2 \times 1$
4	24	$4 \times 3 \times 2 \times 1$
5	120	$5 \times 4 \times 3 \times 2 \times 1$

Worked Example

Question: How many **different arrangements** can be made using the letters **A, B, C, D**?

Steps

Step 1: Use the factorial formula:

$$n! = n \times (n-1) \times (n-2) \times \cdots \times 1$$

Step 2: Identify n:

There are 4 letters, so $n = 4$.

Step 3: Substitute into the formula:

$$4! = 4 \times 3 \times 2 \times 1$$

Step 4: Calculate:

$$4! = 24$$

✓ **Answer:** There are **24** different arrangements of A, B, C, and D.

When to Use Multiplication Principle vs Factorial

Think of counting problems as either **choosing things** or **arranging things**. That helps you decide which method to use.

Use Multiplication Principle (When You Are Choosing)

- ✓ You pick one thing **AND** another thing.
- ✓ Steps are separate.
- ✓ You are selecting, not putting items in order.

If you choose letters separately, you are **choosing**, not arranging.

Choosing = **Multiplication Principle**

Use Factorial (When You Are Arranging)

- ✓ You use all items.
- ✓ Order matters.
- ✓ Items are placed in order (like A, B, C, D).

Arranging = **Factorial**

Checklist

Situation	Method
Choosing letters separately	Multiplication Principle
Arrange A, B, C, D	Factorial
Independent selections	Multiplication Principle
Order of letters matters	Factorial

Remember

- ✓ *Choosing items → Multiply*
- ✓ *Arranging items in order → Factorial*

Avoiding Double Counting

Sometimes items repeat. If we ignore this, identical arrangements are counted more than once. So we adjust for repetition.

Arrangements with Repetition

If n items include repeats with counts $r_1, r_2, \ldots, r_k$:

$$\textbf{Arrangements} = \frac{n!}{r_1!\, r_2! \cdots r_k!}$$

Divide by factorials of repeated items.

Worked Example

Question: How many arrangements of A, A, B?

Think smart:

Without **repetition**:

$$3! = 6$$

Since A **repeats twice**:

$$\frac{3!}{2!} = 3$$

Arrangements: AAB, ABA, BAA.

Grid Path Problems

⚠ Important Reminder

Earlier, grids were used to **locate positions using coordinates**.

Here, grids are used differently:

✕ We are **not locating a point**.

✓ We are **counting how many ways to move across the grid**.

Same grid picture, different purpose.

Movement is usually restricted to:

- Right (R)
- Down (D)

These restrictions:

- Prevent backtracking
- Avoid infinite loops
- Keep the counting process manageable

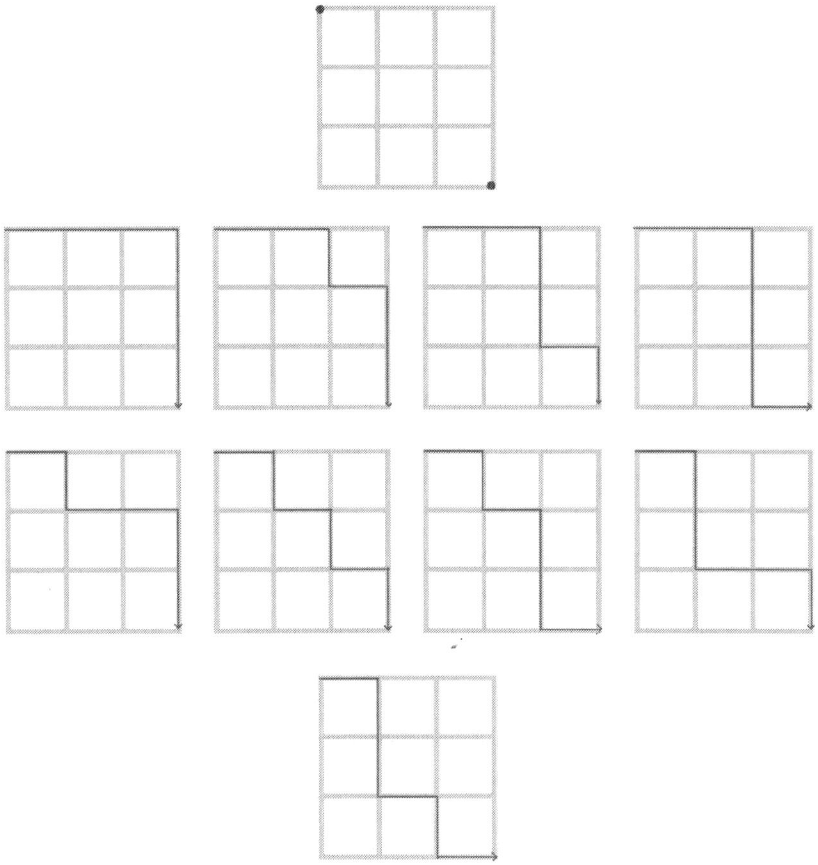

Grid Path Formula

To move across an $\boldsymbol{m} \times \boldsymbol{n}$ grid:

$$\text{Paths} = \frac{(\boldsymbol{m}+\boldsymbol{n})!}{\boldsymbol{m}!\boldsymbol{n}!}$$

Because you are arranging:

- $\boldsymbol{m}$ right moves
- $\boldsymbol{n}$ down moves.

Worked Example

2×2 Grid Moves needed:

$$\text{Move Needed} = \frac{(m+n)!}{m!n!} = \frac{(2+2)!}{2!2!} = \frac{4!}{2!2!} = \frac{24}{2\times2} = \mathbf{6}$$

2×3 Grid Moves needed:

$$\text{Move Needed} = \frac{(m+n)!}{m!n!} = \frac{(2+3)!}{2!3!} = \frac{5!}{2!3!} = \frac{120}{2\times6} = \mathbf{10}$$

Common Traps

- ☒ Listing everything manually (too slow)
- ☒ Forgetting order matters
- ☒ Missing repeated items
- ☒ Ignoring restrictions

Try It Yourself

1. A shop sells 4 hats and 3 jackets. How many outfit combinations?
2. A meal has 2 drinks, 3 mains, 2 desserts. How many meals?
3. How many arrangements of MOM?
4. Paths on a 2×2 grid moving right and down?
5. Arrangements of A, B, C, D?

Key Takeaway

Choose, then choose? Multiply the clues.
Same thing twice? Divide it nice.
Grid moves right and down? Count paths, don't drown.

Choose smart, arrange carefully, avoid double counting.

See the pattern first, the maths becomes easy.

COUNTING METHODS CHEATSHEET: Spot Patterns. Count Smarter.

Focus on structure. Use the right shortcut.

CORE ANCHORS & CHOICES

THE MEMORY ANCHOR (Quick Recall)

× Choose, then choose? **MULTIPLY** the clues.

÷ Same thing twice? **DIVIDE** it nice.

Grid moves R & D? **COUNT PATHS.**

1. MULTIPLICATION PRINCIPLE (Choosing)

Independent choices from separate categories.

Category 1 (n_1) → Category 2 (n_2) → … → Category k (n_k)

$$\text{Total} = n_1 \times n_2 \times \cdots \times n_k$$

ARRANGEMENTS & ORDER

2. FACTORIALS (n!)

(Arranging Distinct Items)

Order matters. Using ALL items.

$$\underline{n} \times \underline{n-1} \times \underline{n-2} \ \ldots$$

$$n! = n \times (n-1) \times \ldots \times 1$$

3. REPETITION (Identical Items)

Divide to avoid double counting identical arrangements.

$$\frac{n! \text{ (Total items)}}{r_1!\,r_2!\ldots \text{ (Factorials of repeats)}}$$

RESTRICTIONS & STRATEGY

4. GRID PATHS (Right & Down Only)

Arranging specific moves (*m* Right, *n* Down).

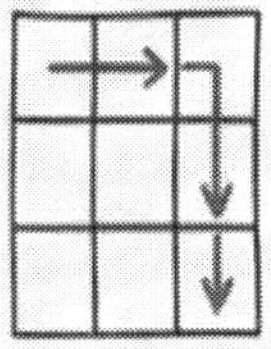

$$\text{Paths} = \frac{(m+n)!}{m!\,n!}$$

5. EXCLUSION METHOD (Restrictions)

Use for 'at least one' or 'not allowed' conditions.

Total Possibilities − Unwanted Cases = Valid Outcomes

DECISION CHECKLIST & TRAPS

Choosing separately? → Multiply.

Arranging in order? → Factorial (n!).

✗ Listing manually (slow)

✗ Ignoring order

✗ Missing repeats

Working Backwards & Reverse Reasoning

Turn complex problems simple by reversing the thinking path.

Most students try to work forward through a problem. Top students often work backwards. When a question looks long, confusing, or full of steps, working backwards can turn it into a one line solution. It is one of the highest value shortcuts in Selective and OC tests.

Your Memory Anchor

Last step first, reverse the way,

Do the opposite, you'll save the day!

Add? Subtract. Multiply? Divide.

Half? Double. Always start from the end.

What Is Working Backwards?

Working backwards means undoing steps in reverse order.
Instead of starting at the beginning, you:

- Start with the final result.
- Reverse each operation one by one.
- Trace back to the starting number.

You're not guessing, you're simply following the path in reverse.

The Three Reverse Rules

Rule 1: Last Step First (Order of Operations)

Always undo the last operation first.

Example:

Forward steps → Multiply by 4 → Add 6
Backward steps → Subtract 6 → Divide by 4

Rule 2: Use the Opposite Operation

Forward Step	Reverse Step
Add (+)	Subtract (−)
Subtract (−)	Add (+)
Multiply (×)	Divide (÷)
Divide (÷)	Multiply (×)
Half	Double
Square	Square root

Rule 3: Trace From the End (Starting Point)

If the question says:

"The final answer is 18..."

Start with **18**, not the unknown number.

Work backwards through each step until you reach the beginning.

When to Use Working Backwards

This is a ***strategy, not a habit***. Use it only when it genuinely helps.

1. The final answer is given

Key phrases:

- ✓ The result is...
- ✓ At the end, the answer is...
- ✓ The final number is...

2. You are asked for the starting number

Look for:

- ✓ What was the original number?
- ✓ Which number did it start with?
- ✓ A machine performs the following operations...

3. The steps are clear and reversible

Example wording:

- ✓ First..., then..., after that...
- ✓ The number is halved, then 4 is added...

4. Forward working becomes messy

If moving forward means algebra, guessing values, or tracking long chains, working backwards is usually faster.

When Not to Use Working Backwards

1. The question asks for the final result

"Start with 6, double it, then subtract 3. What is the answer?"

Just work forward.

2. Steps cannot be reversed

Some actions lose information, such as:

- ☒ Rounding
- ☒ Taking remainders
- ☒ Choosing the largest or smallest value

These cannot be reversed uniquely.

3. There are no clear steps

If it's a direct calculation, pattern question, sequence problem, or geometry task, another strategy may be better.

The 3-Second Decision Test

Ask yourself:

- ✓ Is the final result given?
- ✓ Am I asked for the starting value?
- ✓ Are the steps clear and reversible?

If YES to all three → Work backwards.
If NO to any → Use another method.

End is known, start is asked,
Steps are clear? Reverse the task!

Worked Example

Question: A number **is multiplied by 4**, then **6 is added.** The final **result is 30**. What was the original number?

Think smart:

Remember: Reverse the order and **undo each operation.**

Start at 30
Undo +6

30 – 6 **= 24**

Undo ×4

24 ÷ 4 = **6**

✓ Answer: 6

Question: A number is **halved**, then **3 is subtracted** to give **9**. What was the number?

Think smart:

Remember: Reverse the order and **undo each operation**.

Start at 9
Undo subtraction

$9 + 3 = \mathbf{12}$

Undo halving

$12 \times 2 = \mathbf{24}$

✓ **Answer:** 24

Question: A number is divided by 5, then 4 is added to get 14. Which number did it start with?

A) 30
B) 40
C) 50
D) 60

Think smart:

Remember: Reverse the order and **undo each operation**.

Start at 14
Undo +4

$14 - 4 = \mathbf{10}$

Undo ÷5

$10 \times 5 = \mathbf{50}$

✓ **Answer:** C

Try It Yourself

1. A number is tripled, then 5 is added to get 20. What was the number?
2. A number is divided by 4, then 2 is subtracted to give 6. What was the number?
3. A number is halved, then 7 is added to give 15. What was the number?
4. A number is multiplied by 3, then 10 is subtracted to get 11.

   Which could be the starting number?

   A) 5
   B) 7
   C) 9
   D) 21

5. A number goes through this machine:
   ×2 → −4 → ÷2 → +6 → Final answer: 10. What was the starting number?

Key Takeaway

Last step first, reverse the way,

Do the opposite, you'll save the day!

Add? Subtract. Multiply? Divide.

Half? Double. Always start from the end.

Don't push forward when the answer is already waiting at the end.
Trace backwards. Undo every step. Solve with confidence.

WORKING BACKWARDS & REVERSE REASONING: QUICK REFERENCE CHEATSHEET

Turn complex problems simple by reversing the thinking path. A high-value shortcut.

THE CORE CONCEPT & RULES

WHAT IS IT?

[Final Result] → (Reverse steps one by one) → [Starting Number]

MEMORY ANCHOR:
Last step first, reverse the way,
Do the opposite, you'll save the day!
Always start from the end.

THE 3 REVERSE RULES

1. Last Step First (Order of Operations)
   - Forward: A → B → C
   - Backward: C → B → A
2. Use Opposite Operation

Add (+)	↔	Subtract (-)
Multiply (×)	↔	Divide (÷)
Half	↔	Double
Square	↔	Square Root

3. Trace From the End
   - Start with final answer, not the unknown.

STRATEGY & DECISION MAKING

WHEN TO USE (STRATEGY)

USE WHEN ✓	DON'T USE ✗
✓ Final answer is given	✗ Asked for final result
✓ Asked for starting number	✗ Steps irreversible (rounding, etc.)
✓ Steps clear & reversible	✗ No clear steps (patterns, geometry)
✓ Forward working is messy	

THE 3-SECOND DECISION TEST

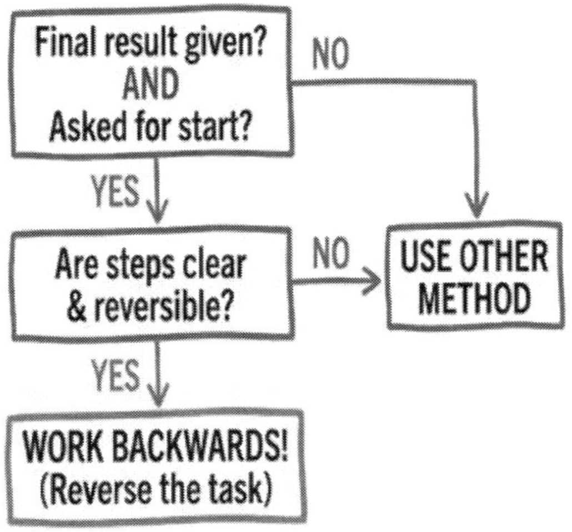

KEY TAKEAWAY & SUMMARY

CORE PHILOSOPHY

- Don't push forward when the answer is already at the end.
- Trace backward. Undo every step.
- Solve with confidence.

Last step first,
reverse the way,
Do the opposite, you'll
save the day!

Add? Subtract.
Multiply? Divide.
Half? Double.

Always start from the end.

Common Mistakes (and How to Avoid Them)

Understanding Exam Traps to Improve Accuracy and Confidence

Most incorrect answers in selective school exams do not happen because students lack knowledge. They happen because of rushing, misreading, skipping logical steps, or choosing speed over clarity. Selective exams are designed deliberately to test attention, interpretation, reasoning, and composure under pressure.

High-performing students are not those who never make mistakes. They are the ones who recognise common traps early and respond strategically. This chapter focuses on those traps — how they appear, why they cause confusion, and how to avoid them.

Think of this as your mistake-awareness guide rather than a mathematics lesson.

Misreading What the Question Really Asks

Example Question

A cyclist travels 210 km in 3 hours. Find the average speed in metres per second.

Where the Trap Is

Students often calculate correctly in km/h but **overlook the required** unit at the end.

How to Avoid It

- ✓ Always read the final line carefully before calculating.
- ✓ Identify exactly what form the answer must take.
- ✓ Underline units or keywords if helpful.

Hidden Multi-Step Reasoning

Example Question

A rectangle has a perimeter of 56 cm. Its length is 6 cm more than its width. Find its area.

Where the Trap Is

Students sometimes find the dimensions but forget the **final requirement**.

How to Avoid It

- ✓ Check whether your answer is intermediate or final.
- ✓ Ask: "Have I answered the actual question?"

- ✓ Expect multiple steps in selective exam problems.

Distractor Information

Example Question

A school has 720 students. 65% participate in extracurricular activities. On one day, 18 students are absent. How many students participate?

Where the Trap Is

The absence information is **irrelevant** but included to distract attention.

How to Avoid It

- ✓ Focus on what directly relates to the question.
- ✓ Do not assume every number must be used.
- ✓ Simplify mentally before calculating.

Language Designed to Mislead

Example Question

A price was reduced to 75% of its original value.

Where the Trap Is

Students confuse "reduced to" with "reduced by," which **implies a different operation**.

How to Avoid It

- ✓ Translate wording into mathematical meaning first.
- ✓ Read slowly when wording seems subtle.
- ✓ Never assume familiar phrasing means the same thing.

Skipping Logical Steps

Example Question

A complex algebra expression simplifies to a numerical value.

Where the Trap Is

Students sometimes perform mental shortcuts and make **sign, unit, or arithmetic errors**.

How to Avoid It

- ✓ Write essential steps clearly.
- ✓ Avoid doing everything mentally under time pressure.
- ✓ Accuracy usually improves with structured working.

Pattern Recognition Pitfalls

Example Question

3, 9, 10, 30, 31, 93, ... What comes next?

Where the Trap Is

The sequence alternates operations, but students often assume a single consistent pattern.

How to Avoid It

- ✓ Check multiple terms before deciding.
- ✓ Consider alternating or layered patterns.
- ✓ Avoid guessing based on appearance.

Diagram and Graph Interpretation Traps

Example Question

A bar graph shows student participation where each grid line represents five students.

Where the Trap Is

Students may count bars visually **without checking the scale**.

Similarly, diagrams may not be drawn to scale.

How to Avoid It

Always check:

- ✓ titles
- ✓ axes
- ✓ measurement labels
- ✓ scale information

Trust data, not appearance.

Rushing and Time Pressure Decisions

Example Scenario

An unfamiliar question appears early in the exam.

Where the Trap Is

Students may spend **excessive time trying** to solve it immediately, reducing time for easier questions.

How to Avoid It

- ✓ Move on temporarily if progress stalls.
- ✓ Maintain steady pacing.
- ✓ Return later with a clearer mind.

Strategic skipping is often a strength, not a weakness.

Unit and Scale Awareness

Example Question

Convert an area measurement from square metres to square centimetres.

Where the Trap Is

Students convert linear units correctly but forget squared or cubed scaling.

How to Avoid It

- ✓ Pay special attention to area and volume units.
- ✓ Check consistency of units before finalising answers.
- ✓ Confirm whether conversion applies once, squared, or cubed.

Logical Reasoning Oversights

Example Question

A probability question involves multiple possible outcomes.

Where the Trap Is

Students sometimes guess intuitively rather than considering all possibilities.

How to Avoid It

- ✓ Pause to consider logical constraints.
- ✓ Ask whether the answer is realistic.
- ✓ Balance calculation with reasoning.

Reality-Check Thinking

Example Situations

- Probability greater than 1
- Impossible measurements
- Percentages exceeding logical limits

Where the Trap Is

Correct calculation can still produce unrealistic answers if **interpretation is wrong**.

How to Avoid It

Before moving on, ask:

- ✓ Does this answer make sense?
- ✓ Is it physically or mathematically possible?
- ✓ Would estimation give a similar result?

This habit prevents many careless errors.

Smart Guessing and Avoiding Blank Answers

Example Scenario

Time is nearly finished and one question remains.

Where the Trap Is

Students panic and **leave the question blank**.

Most selective tests do not penalise incorrect answers.

How to Avoid It

- ✓ Eliminate clearly incorrect options first.
- ✓ Make an informed attempt rather than guessing randomly.
- ✓ Never leave a question unanswered if guessing is allowed.

The Selective Exam Self-Check Habit

Before moving on from any question, briefly confirm:

- Did I answer exactly what was asked?
- Are units correct?
- Did I overlook any condition?
- Does the answer seem reasonable?

This short review significantly improves accuracy.

Key Takeaway

Selective school exams reward disciplined thinkers more than fast calculators.

Students who consistently perform well tend to:

- read carefully
- interpret precisely
- manage time strategically
- remain calm under pressure
- check answers thoughtfully

Mistakes are normal. Learning how to recognise and prevent them is what leads to improvement. Accuracy grows from awareness, and awareness grows from practice.

Cool Maths Tricks (Because Your Brain Deserves Fun!)

Secret toolbox of selective school champions!

By now you've tackled:

- ✓ Challenging maths problems.
- ✓ Tricky reasoning questions.
- ✓ Exam-style traps.
- ✓ Lots of practice.

So, this chapter is different.

This is your:

- ✓ Maths fun zone.
- ✓ Brain refresher.
- ✓ Confidence booster.

These aren't always essential for exams, but they make maths quicker, smarter, and way more enjoyable.

Think of them as your **legal cheat codes for maths success.**

Ready? Let's go.

Speed Conversion Trick (km/h → m/s)

Sometimes speed is given in **km/h** but needed in **m/s**.

Speed in m/s = Speed in km/h ÷ 3.6

Example

72 km/h ÷ 3.6 = **20 m/s**

The ×11 Lightning Trick

One-digit numbers:

Just repeat the number:

4 × 11 = 44
7 × 11 = 77

Two-digit numbers:

Add digits and place sum in the middle.

35 × 11
3 + 5 = 8
✓ 385

If the sum is bigger than 9, carry:

76 × 11
7 + 6 = 13
✓ 836

The 9 Times Finger Trick

Try this:

1. Hold up both hands
2. Bend the 7th finger (for 9×7)
3. Fingers before = tens
4. Fingers after = ones

✓ 63

Squaring Numbers Ending in 5

- ✓ Multiply first digit(s) by the next number.
- ✓ Add **25** at the end.

Example:

35^2
3 × 4 = 12
✓ 1225

65^2 → **✓ 4225**

The Rule of 6

Even number × 6?

- ✓ Ones digit stays the same.

Example:

6 × 14 = 84
Same ones digit = 4.

The Rule of 5

- ✓ Divide the number by 2.
- ✓ Multiply the result by 10 (add a zero).

Example: 5 × 8

Step 1:

8 ÷ 2 = 4

Step 2:

4 × 10 = 40

Multiply by 9 Fast

- ✓ Multiply by 10.
- ✓ Subtract the number.

Example: 9 × 6

= (6 × **10**) – **6 =** 60 – 6 **= 54**

Double-Double Trick (×4)

Multiply by 4:

- ✓ Double
- ✓ Double again

Example: 4 × **9**

9 → 18 → **36**

Multiply by 99 Trick

- ✓ Multiply by 100.
- ✓ Subtract original number.

Example: 67 × 99

= (67 × **100**) – **67** = 6700 – 67 = **6633**

Memorising Pi

Remember: 3.141592

Sentence: *"How I wish I could calculate pi."*

Count letters:

3-1-4-1-5-9-2

Divisibility Detective Tricks

Number	**Quick Test**
2	Ends 0,2,4,6,8
3	Digit sum divisible by 3
4	Last 2 digits ÷ 4
5	Ends 0 or 5
6	Even AND digit sum ÷ 3
8	Last 3 digits ÷ 8
9	Digit sum ÷ 9
10	Ends 0

Percentage Power Trick

Finding 10% of 320

Move the decimal one place to the left.

10% of 320 = ✔ 32

Finding 5% of 320

First find 10%, then take half of it.

10% of 320 = 32

Half of 32 = **16**

Number Magic Trick (Answer Always 2!)

Try:

1. Think of a number
2. ×3
3. +6
4. ÷3
5. Subtract original number

Answer = **2 every time.**

Estimation = Secret Weapon

Before solving, ask:

- ✓ What should this roughly be?

Example:

$$198 + 302 \approx 200 + 300 = 500$$

Top students estimate first.

Counting Numbers by Digits (Easy Trick)

Last number – First number + 1

Use this to count how many numbers are in any range.

Example: 1 to 150

Digits	Range	Count
1-digit	1–9	9
2-digit	10–99	99 – 10 + 1 = 90
3-digit	100–150	150 – 100 + 1 = 51

Total numbers = 9 + 90 + 51 **=** 150

Example: 1 to 500

Digits	Range	Count
1-digit	1–9	9
2-digit	10–99	99 – 10 + 1 = 90
3-digit	100–500	500 – 100 + 1 = 401

Total numbers = 9 + 90 + 401 **=** 500

Your Maths Exam Quick-Recall Tables

All in one place before test day.

This chapter isn't about learning new maths.
It's about **quick recall, confidence, and avoiding common exam traps.**

Before any maths exam, especially Selective or OC, students benefit from:

- ✓ Recognising key maths language quickly
- ✓ Remembering essential facts automatically
- ✓ Checking common mistakes calmly
- ✓ Thinking clearly under time pressure

Use these tables as your **final mental refresh before stepping forward.**

Maths Language Decoder

Term	What It Means	Example
Sum	Addition	Sum of 9 and 4 = 13
Difference	Subtraction (larger − smaller)	Difference between 18 and 7 = 11
Product	Multiplication	Product of 8 and 5 = 40
Quotient	Division	24 ÷ 6 = 4
Factor	Divides exactly (no remainder)	4 is not a factor of 18
Multiple	Result of multiplying	Multiples of 6: 6, 12, 18…
Of	Usually means multiply	$\frac{3}{5}$ of 25 = 15
Each / Per	Repeated groups	7 bags × 6 apples = 42
Shared equally	Divide into equal parts	24 ÷ 6 = 4
Total / Altogether	Combine amounts (add)	9 + 6 = 15
Remaining / Left	Subtract what was used	30 − 12 = 18
At least	Minimum (≥)	At least 20 → 20 or more
At most	Maximum (≤)	At most 50

Number & Operations Essentials

Skill	Key Reminder	Example
Column addition	Line up place values	368 + 47 = 415
Subtraction	Borrow carefully	500 − 268 = 232
Split multiplication	Break into easier parts	24×6 = 120 + 24
Division check	Multiply to verify	56 ÷ 7 = 8
Remainder rule	Dividend = divisor × quotient + remainder	17 = 5×3 + 2
Estimation	Round first to check answers	198×6 ≈ 1200
Place value awareness	Digit value depends on position	30,000 vs 60 → 500× larger

Fractions, Decimals & Percentages

Idea	Key Reminder	Example
Fraction of number	Multiply	⅖ of 30 = 12
Percent meaning	Out of 100	25% = ¼
Decimal ↔ Fraction	Same value, different form	0.6 = 3/5
Benchmarks	Memorise common values	½, ¼, ¾
Comparing fractions	Same denominator → bigger numerator wins	5/8 > 3/8
Percent calculation	Percent × number	15% of 60 = 9
Probability link	Fraction can be written as percent	¼ = 25%

Units & Measurement

Topic	Key Idea	Example
Length	Convert units first	2.3 m = 230 cm
Mass	1000 g = 1 kg	1500 g = 1.5 kg
Capacity	1000 mL = 1 L	1.2 L = 1200 mL
Time	Use one unit consistently	2h 45m = 165 min
Area	Uses square units	5×3 = 15 cm^2

Volume	Uses cubic units	2×3×4 = 24
Special link	1 cm^3 = 1 mL	250 cm^3 = 250 mL
Speed	Keep units consistent	km/h or m/s

Geometry Essentials

Concept	**Key Fact**	**Example**
Right angle	90°	Corner of a square
Straight angle	180°	Half turn
Full turn	360°	Complete rotation
Triangle angle sum	Always 180°	50°+60° → 70°
Rectangle area	Length × width	12×4 = 48
Rectangle perimeter	2(l+w)	8×5 → 26
Composite shapes	Split into smaller shapes	Add areas
Symmetry	Mirror line halves match	Square has several
Diameter	Twice the radius	r=6 → d=12

Data & Probability

Idea	**Meaning**	**Example**
Mean	Average	4,6,8 → 6
Median	Middle value (ordered list)	1,2,5,7,9 → 5
Mode	Most frequent value	4 appears most
Range	Largest – smallest	11−2 = 9
Probability	Favourable ÷ total outcomes	3/12 = ¼
Missing probability	Total probability = 1	1−0.75 = 0.25
Graph reading	Always check scale	Each gap = 5

Patterns & Reasoning Skills

Skill	**What To Remember**	**Example**
Number patterns	Look for + − × ÷ first	3,6,12,24 → ×2
Shape patterns	Observe growth/change	Add 3 tiles each step
Balance problems	Equal totals both sides	2 blocks = 10 kg
Working backwards	Reverse operations	Undo × then +
Elimination	Remove impossible answers	Negative area impossible

Final Test-Day Checklist

This Is Your Game Plan for Success

By now, you've learned:

- ✓ Smart methods
- ✓ Time-saving tricks
- ✓ How to avoid common mistakes

This final chapter brings it all together into a simple, calm plan for test day.
No stress. No cramming. Just clear thinking and confidence.

The Night Before the Test

Review the essentials

- ✓ Fraction–decimal–percentage facts.
- ✓ Prime numbers.
- ✓ Angle rules.
- ✓ Unit conversions.
- ✓ Key formulas (area, perimeter, probability).

Do a few easy questions
Not a full test, just enough to:

- Stay sharp.
- Build confidence by doing a few easy questions, not a full test, just enough to stay sharp.

Pack what you need

- Pencils/pens.
- Eraser.
- Water bottle.
- Any required documents.

Get a good night's sleep

- ✓ Your brain works best when it's rested.

Don't Do These

- ☒ Don't try to learn brand-new topics.
- ☒ Don't stay up late "just in case."
- ☒ Don't panic over one mistake in practice.

Trust the work you've already done.

The Morning of the Test

Do These

- ✓ Eat a simple breakfast.
- ✓ Arrive early so you're not rushed.
- ✓ Take a few deep breaths.
- ✓ Remind yourself:

I don't need to know everything. I just need to think clearly.

In the Test Room

Your Step-by-Step Strategy

Step 1: Start Smart

- ✓ Read the instructions calmly.
- ✓ Don't rush the first question.
- ✓ Find your rhythm.

Step 2: Use the Easy-First Rule

Begin with questions that:

- ✓ Look familiar
- ✓ Have short working
- ✓ Feel clear after one read

Leave harder ones for later.
Easy marks first, always.

Step 3: Use the 1-Minute Rule

If you:

- ☒ Feel stuck
- ☒ Can't see a clear method

✔ **Skip it and move on**. You can come back later.

Step 4: Read Carefully

Before solving:

- ✓ Read the question twice.
- ✓ Look for keywords like:
  - difference
  - total
  - least / greatest
  - of
  - closest to

Many mistakes come from misreading, not misunderstanding.

Step 5: Use Multiple Choice to Your Advantage

You can:

- ✓ Eliminate impossible answers
- ✓ Estimate
- ✓ Substitute options back into the question

Multiple choice is a tool, not a trap.

Step 6: Think Smart, Not Long

Ask yourself:

- ✓ Is there a shortcut?
- ✓ Can I divide first?
- ✓ Can I estimate?
- ✓ Can I work backwards?

Use the smart methods you've learned.

The Last 5 Minutes

This is where you gain extra marks.

Do This

- ✓ Return to skipped questions.
- ✓ Check:
  - units
  - signs (+, −, ×, ÷)
  - rounding
- ✓ Make sure every question has an answer.

Don't Do This

- ☒ Don't change answers without a clear reason.
- ☒ Don't panic.
- ☒ Don't leave any blanks.

There is no penalty for guessing.

Your Test-Day Mantra

Say this quietly to yourself:

"Read carefully. Think clearly. Move on when stuck. Try every question."

You don't need to be perfect.
You just need to be calm, smart, and focused.

Checklist

The Night Before

- ✓ Light revision
- ✓ Pack your things
- ✓ Sleep well

The Morning

- ✓ Eat
- ✓ Arrive early
- ✓ Breathe

During the Test

- ✓ Easy first
- ✓ 1-minute rule
- ✓ Eliminate
- ✓ Estimate
- ✓ Stay calm

The End

- ✓ Check units
- ✓ Try skipped questions
- ✓ Answer everything

Final Words

You have learned how to:

- ✓ Think, not memorise
- ✓ Use smart shortcuts
- ✓ Avoid common traps
- ✓ Stay calm under pressure

This test is not about being a genius.
It's about clear thinking, good habits, and confidence.

💡 Remember

You are ready. You are capable. You can do this.

Test Strategy (Quick Summary)

- ✓ ~1 minute per question
- ✓ Easy first, hard later
- ✓ Skip and return if stuck
- ✓ Eliminate, estimate, substitute
- ✓ Never leave blanks

Read carefully. Think clearly. Move on when stuck. Try every question.

A Final Word Before You Step Forward

First, congratulations.
If you are reading this, you have come a long way. Finishing this book is already an achievement. It shows commitment, patience, and the willingness to improve, and that matters far more than natural talent.

Remember this always: nobody is born a genius. People become confident, capable, and successful through consistent effort, practice, mistakes, learning, and the courage to keep going right up to exam day.

Your hard work deserves recognition. So does the support system behind you, especially your parents or guardians who stood beside you with encouragement, belief, and unconditional support. Success is rarely a solo journey, and their role is part of your story too.

As you step into your test:

- ✓ Trust your preparation.
- ✓ Stay calm.
- ✓ Think clearly.
- ✓ Do your best. That is always enough.

Marks matter, but growth matters more. The habits you have built here, including persistence, smart thinking, and confidence under pressure, will serve you far beyond any single test.

Make this moment memorable not because it is the end of a book, but because it marks the beginning of believing in what you can achieve through effort.

You have done the work.
Now go show it.

Good luck. You have got this.

Made in United States
Orlando, FL
04 November 2026

90218341R00136